Invertebrate Zoology: A Laboratory Manual

By Alan R. Holyoak

Text copyright © 2013 Alan R. Holyoak

Images copyright (except as otherwise credited in the text) © 2013 Alan R. Holyoak

All Rights Reserved

The author retains the right to make copies of the work available for internal distribution by Brigham Young University-Idaho

Dedication

To Todd Newberry, John Pearse, Mike Hadfield, Lee Braithwaite, and Larry Hibbert, mentors and friends who fed my thirst for knowledge about all things invertebrate, and most of all to my wife Kathrine for happily joining me in our life-long biological adventure

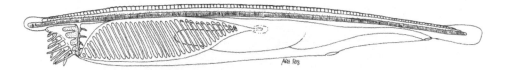

Alan Holyoak earned BS and MS degrees in Zoology from Brigham Young University, and a PhD in Biology from the University of California, Santa Cruz. He has taught invertebrate zoology for over 20 years, first at Manchester University (Indiana) and currently at Brigham Young University-Idaho. During his career he has also taught or carried out research at The Friday Harbor Laboratories of the University of Washington, The University of Hawaii, The Long Marine Laboratory of UC Santa Cruz, The Hopkins Marine Station of Stanford University, and The Oregon Institute of Marine Biology of the University of Oregon.

Table of Contents

PREFACE	6
INTRODUCTION	7
THE LABORATORY NOTEBOOK	8
CHAPTER 1: INTRODUCTION TO MICROSCOPY	12
CHAPTER 2: PHYLOGENETIC ANALYSIS	18
CHAPTER 3: SPONGES	25
CHAPTER 4: CNIDARIA	32
CHAPTER 5: PLATYHELMINTHES	42
CHAPTER 6: MOLLUSCA	51
CHAPTER 7: ANNELIDA	72
CHAPTER 8: BRACHIOPODA AND NEMATODA	81
CHAPTER 9: PANARTHROPODA	90
CHAPTER 10: ECHINODERMATA	107
CHAPTER 11: HEMICHORDATA AND CHORDATA	118
REFERENCE MATERIAL	123
INDEX	124

Preface

The price of textbooks marketed by traditional publishing houses has spiraled out of control. At the time of this writing new copies of invertebrate zoology textbooks cost up to $170 and laboratory manuals up to $85. It frankly pains me to require students to shell out that kind of cash for one class. I therefore decided to produce a laboratory manual that focuses specifically on the experiences I want my students to have, and simultaneously helps alleviate some of the financial stress they face.

This manual is designed to support an upper division, undergraduate course in invertebrate zoology. Students who use these exercises will gain the greatest benefit from them if they have already completed a course in which they were introduced to the unifying principles of biology, including ecology and evolution, before taking invertebrate zoology.

My teaching philosophy includes the idea that everyone has something to share, and that everyone has something to learn. While you can complete the exercises in this manual individually, I strongly encourage you to work in pairs or even in small groups. This way you always have someone you can share ideas, observations, and questions with. I have observed that just about everyone gets more out of laboratory experiences when they share what they learn as they learn. Each chapter also includes at least one "group question" that should be discussed and answered in small groups.

I strongly encourage you to produce a laboratory notebook as part of your laboratory experience. Examining specimens, drawing them, and then writing down observations and questions about them will help you develop important observational and record keeping skills. Your lab notebook is also a permanent record of what you accomplished in lab, and in many cases becomes a treasured memento.

Lastly, the species used in these exercises can usually be obtained from biological supply companies. This makes it possible to carry out the majority of exercises in this manual anywhere, either coastal or inland. You will, I hope, also study locally available species whenever possible. I hope this manual is helpful to you as you study and teach invertebrate zoology.

Alan R. Holyoak
Department of Biology,
Brigham Young University-Idaho

Introduction

Welcome to Invertebrate Zoology! Invertebrate zoology is where it's at if you are interested in the diversity of animals, their body plans, and the strategies they use to survive and reproduce. This manual outlines exercises in which you can compare and contrast the diversity of animal body plans ranging from anatomically simple animals, such as sponges, to our own taxon, Chordata.

Invertebrate Zoology is also a fantastic discipline for learning applied laboratory skills including microscopy, dissections, producing laboratory drawings, and developing your scientific thinking skills of as you make observations and ask questions about specimens you study.

We are all learners and teachers, so share what you see and learn with each other as you work. There will therefore be times you need help. When those times come, ask! There will also be times when you can help someone else. When those times come, don't wait to be asked! Lastly, your ability to think and learn as a scientist grows as you actively pursue knowledge and understanding. When you have an idea or a question, great or small, jot it down and then share it. Remember that the acquisition of knowledge is only the accumulation of information and by itself doesn't have much impact, but true learning is measured by the impact that knowledge has on you. That is, learning not only changes what you know, it changes what you do.

The exercises in this laboratory manual are designed to get you looking at and thinking about invertebrate animals. I encourage you to move beyond the information in this manual and expand your knowledge by making additional observations, producing additional drawings, and seeking deeper understanding from each body plan you encounter in lab. Remember that each animal body plan is ecologically and evolutionarily successful or it wouldn't still be around. Our challenge and opportunity as invertebrate zoologists is to discover the ecological and evolutionary solutions each body plan presents.

I hope you will work as a community of scholars – a gang of budding invertebrate zoologists. Being a member of a community of scholars means that you must to prepare for each lab meeting before you arrive to do the hands-on work. Read through the lab materials prior to showing up for lab meetings. This way you can maximize the time available to do the fun stuff, hands-on work.

You will note that there is a set of "Group Questions" at the end of each chapter. Discuss and answer these questions in small groups. In order to answer these questions you need to rely on a combination of experience from the lab, information from class, and sometimes even additional material that you need to track down from outside sources.

Enjoy the adventure!

The Laboratory Notebook

One of the main goals of laboratory work is to give you opportunities for hands-on experiences that help you learn to work and think as scientists. An important part of this work is keeping a complete and accurate record of what you do in lab. I cannot overstate the importance of keeping accurate records and developing a well-formatted notebook. The ability to do these things will yield great dividends when you begin carrying out your own research. This section includes ideas about how you can develop a laboratory notebook, at least what I require my students to do.

These are some things you will need to bring to lab. Use a 7" x 10" or 9" x 12" hardcover spiral bound artist's sketchbook for your laboratory notebook. The cover protects your work throughout the course and beyond, and the spiral binding allows the notebook to lie flat while you work. This comes in handy while you are observing something and drawing it or jotting down notes about it at the same time. You also need 2H or 3H pencils. The lead in regular #2 (HB) pencils is too soft and will smear over time, while 2H-3H pencils produce a nice line and have lead that is hard enough that doesn't easily smear. You should also have an artist's kneaded eraser as well as a ruler with metric divisions.

Notebook formatting

1. Contact information: Write your name, phone number, and email address on the inside of the front cover of your notebook. You will put a lot of work into your laboratory notebook, and this will increase the chances of it being returned to you if, perish the thought, it gets lost or stolen.
2. Microscope ocular micrometer calibrations or aperture field width measurements: Write the ocular micrometer measurements for the microscope you are using inside the front cover of your notebook. If your microscopes don't have ocular micrometers in them you can still use a stage micrometer to measure the field widths of the field of view for each magnification of a compound microscope. You need this information because you will use it to generate scale bars to accompany some drawings you produce.
3. Table of Contents: Leave the first two or three pages of your notebook blank when you make your first entry. Use these pages for your Table of Contents. A Table of Contents is essential to quickly and easily finding what you are looking for in your notebook. Each entry in your Table of Contents should include a page title, page number, and the date the work was done.
4. Page formatting for the notebook: Write the page number, date, and page title at the top of every page of your notebook. A page title is just a brief description of what's on the page. It can be as simple as "Mollusc Lab #1" or a description of whatever is on the page.

Laboratory Drawings

Why produce drawings? Most people roam the planet looking at lots of things, but seeing very little, e.g., they look around and see "a house", "a tree", "a person", "a dog", etc., but don't go any farther than that. Artists take this one step farther. They are trained to actually see what they look at. They look at "a tree" and they see patterns of light and dark, the texture of the bark, the angles and patterns of branching, the sizes and shapes of leaves, etc. Scientists take this one step farther. Scientists strive to see what they are looking at, and then describe and explain what they see. They do this by making observations and asking questions. A scientist may ask, "Is branching pattern of this tree adaptive, and if so, how?" Scientists then collect observations and use them to try to answer their questions. Producing laboratory drawings, and you should produce a lot, will help you develop observational skills of an artist and the curiosity of a scientist. Your work pattern during lab should be to look, see, and then ask questions. As you practice this pattern of observing, drawing, and asking you will develop observational and question-asking skills used by scientists.

Drawing also forces you to look for detail. It has been said that a pencil is a great aid to observation. Most of the drawings you will produce are assigned in the body of laboratory exercises, but in order to gain full benefit from your lab experience you should push yourself to produce and include additional drawings of what you see. Remember, the goal of producing lab drawings is to help you observe you are looking at, as well as to provide a record of what you did during labs.

Lab drawings are working drawings and as such need not be "artistically beautiful". Working drawings help you recall what you saw and did in lab. To that end, your drawings should represent what YOU saw, not what is shown in drawings or photographs in the lab exercises or in the textbook. Draw what YOU see, don't just replicate a drawing from the lab manual. The entries in your lab notebook should provide you with a good enough record of your experiences so that if the occasion requires you can turn to your lab notebook for review rather than cutting up or observing another specimen.

Guidelines for lab drawings

1. Draw big! Fill at least half of the page with one drawing. In most cases have only one drawing per page. This is important, because the larger a drawing is, the more detail you are likely to include in it. Drawing large forces your eye to look for detail. Try to include everything you see in your drawings, even though you may not know what everything is. This will, in turn, lead you to asking better questions.
2. Each of your drawings needs a scale bar. The next section explains how to generate a scale bar.
3. Begin each drawing by including the most obvious things, and then add details as you develop a drawing.
4. Don't hesitate to include more than one drawing from a single dissection. In fact, the more drawings you make, and the more detailed they are, the more you will see, and the better you will know each specimen.
5. Always draw in pencil. Use either a 2H or a 3H pencil. Never use pen in your notebook!
6. Do not render (i.e., shade or color) your drawings. Shading or coloring often obscures detail rather than enhances it. Instead, describe the things you see via accompanying observations.
7. Feel free to use drawing aids as you see fit, i.e., rulers, circle templates, compasses, calipers, etc. Do not expect drawing aids to be provided. Some of my students have taken to photographing specimens, but photographs cannot replace hand-produced lab drawings for developing observational skills, and do not replace required laboratory drawings when it is time to grade lab notebooks.

How to Generate a Scale Bar

Every drawing in your lab notebook needs a scale bar. A scale bar is an indication of the size of your drawing in relation to the actual size of the specimen you drew. The easiest way to explain how to generate a scale bar is to have you work through an example. Look at Figure 0.1. It shows a photograph of a specimen and a drawing of that specimen.

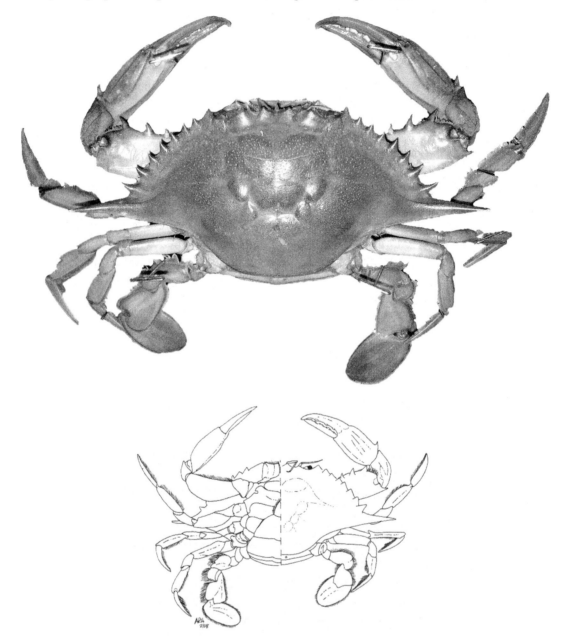

Figure 0.1. Photograph of a specimen (above) and drawing of the specimen showing dorsal and ventral anatomy (below).

To generate a scale bar measure some linear distance on the original specimen and the same linear distance on your drawing. Divide the linear distance on your drawing by the linear distance on the original specimen. This gives you a numberless value.

Task – generate a scale bar for the drawing in Figure 0.1

1) Measure the distance from the tip of the large spike-like spine on the left side of the carapace of the photograph of the crab to the tip of the large spike-like spine on the right side of the carapace. Use the metric scale for your measurements.
2) Measure the distance between tips of the spike-like spines on the drawing.
3) Divide the distance from the drawing by the distance from the photograph.
4) If you did everything correctly you should get a quotient of about 0.53. This value represents the size ratio between the specimen and the drawing, 0.53 to 1.0. Check with your neighbor or the instructor to see where you missed something if you didn't get 0.53 or close to it.
5) Now that you have your scaling value, multiply it by some convenient linear distance, e.g., 1.0 cm or 5.0 cm or 10.0 cm. If you choose 10.0 cm you will get a product of 5.3 cm. Use a ruler and draw a line 5.3 cm long just below your drawing, and write "10.0 cm" above the line. This tells the reader that the line you drew represents 10.0 cm on the original specimen. Your scale bar is complete.

Observations and Questions

You are not done with an entry when you complete a drawing and its scale bar. You still need to label all the structures that you can on your drawing, and then write down at least three observations and three questions pertaining to each entry you make in your lab notebook. Observations can clarify what you see as you develop a drawing or they can be thoughts you have about your specimen as you study it. Questions can be about any aspect of the specimen you are observing. They will, I hope, be insightful and often lead you to additional questions and observations.

I recommend asking questions that start with "I wonder how/if/what/why/when..." if you don't know where to start when it comes to posing questions. Also, any question you would ask your neighbor or the instructor about your specimen could and perhaps should be jotted down. And, you can certainly have more than three questions per entry.

I recommend having a designated space on each page for a drawing, for observations, and for questions. One approach is to draw a line 6 to 8 cm above the bottom of a page, and then divide the area below the line in half. Use the large area above the line for your drawing, one of the smaller areas below the line for your observations and the other one for your questions. When you do this on every page, it helps you remember that you need observations and questions for each entry.

Chapter 1: Introduction to Microscopy

This laboratory exercise introduces you to principles of good microscopy, and familiarizes you with the anatomy of compound and dissection microscopes. About now you might be thinking, "Why do I need an introduction to microscopy? I already know how to use a microscope." I don't doubt that you have had lots of chances to use microscopes, but in my experience too few students receive adequately instruction in microscopy. And, because you are now clearly on the path to being scientists it is time that you learned the finer points of microscopy.

The compound microscope:

You use a compound microscope to look at specimens that are small enough, thin enough, or transparent enough to allow light to pass through them. This microscope has a light source in the base that projects light up through a specimen and into the lens system.

Task #1: Use Figure 1.1 and descriptions following it to become familiar with the parts of a compound microscope

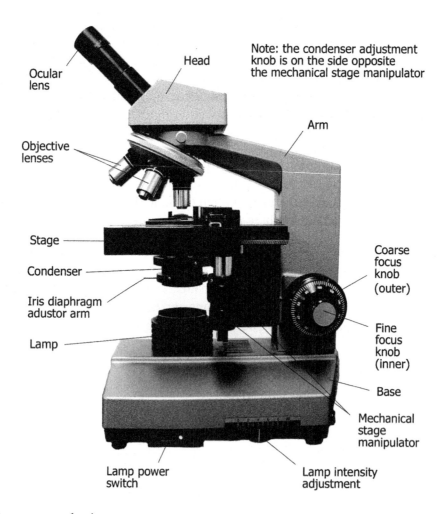

Figure 1.1. A compound microscope.

Parts of the compound microscope

Ocular lens or eyepiece – typically a 10x lens that slides into the barrel of the head.

Objective lenses – lenses of different magnifications mounted on a rotating base located on the base of the head. Look at the objective lenses to see their magnification powers.

Stage (with spring-loaded slide holder) – supports the slide during observation.

Mechanical stage manipulator – dual rotating knobs move the slide forward and backward and side-to-side across the stage.

Condenser – located beneath the stage, focuses light from the lamp to maximize resolution

Iris diaphragm adjustor arm– this lever opens and closes the **iris diaphragm**, a dilating structure mounted on the condenser. The iris diaphragm regulates the amount of light passing from the lamp to the objective lens. Closing the iris diaphragm increases the contrast of the image: opening the diaphragm increases image brightness. Optimal image quality is obtained by adjusting both the lamp intensity and iris diaphragm.

Coarse/fine focus knobs – the outer knob is the coarse focus. Use this knob to find the focal plane for each magnification. The inner knob is the fine focus. Use this knob to fine-tune the focus once you have found the focal plane.

Lamp power switch – switches the lamp on and off.

Lamp intensity adjustment—this sliding lever is used to adjust the amount of light produced by the lamp. There is a tendency among less experienced microscopists to set the lamp intensity too high. This washes out the image, fatigues the eyes, and can cause headaches. Increase lamp intensity only enough to see the image clearly.

Condenser adjustment – this knob moves the condenser up and down to achieve optimal focusing of the light from the lamp.

Arm – supports the head

Head—houses mirrors that reflect the incoming light from the objective lens to the ocular lens. There is usually a thumbscrew that can be loosened to allow the head to swivel to a preferred position. Once the head is in position, however, make sure the thumbscrew is re-tightened. If the thumbscrew is not tight the head could fall off while the microscope is being moved. The head also supports the barrel for the ocular lens, as well as the rotating support for the objective lenses.

Base – supports the scope and houses the lamp.

Lamp—Note the notch in the front of the lamp housing. This comes into play when you adjust the condenser, as described later in this chapter.

The Dissection Microscope

A dissection microscope is used for observing specimens that are too large or too opaque for light to pass through them. Light reflects off of the surface of these kinds of specimens and up into the microscope.

Task #2: Use Figure 1.2 and the following descriptions of parts to become familiar with the dissection microscope.

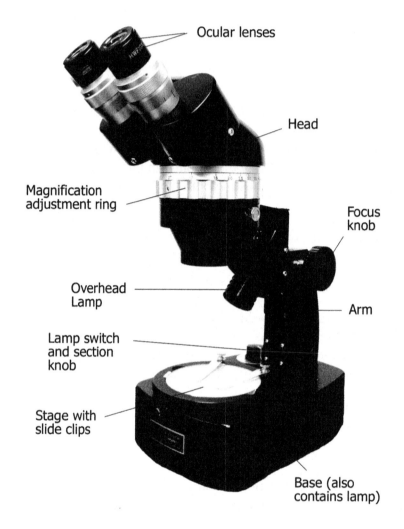

Figure 1.2. A dissection microscope.

Parts of the dissection microscope

Ocular lenses or eyepieces – same as on the compound scope (10x mag.)

Head, Arm, and Base – same as on the compound scope

Magnification adjustment ring – this rotating ring provides a magnification zoom capability, the same general role as objective lenses on the compound scope. The range of magnification on this dissection scope is 0.7x to 4.5x. So together with the ocular lens, its range of magnification is 7x to 45x.

Focus knob – used to adjust focus.

Overhead lamp – projects light down onto the specimen.

Stage – supports the specimen during observation. Stage clips hold slides in place. Stage clips are easily removed, as needed. The centerpiece of the stage can be opaque white, opaque black, translucent, or clear glass, and can be changed as needed.

Lamp switch and selection knob – the dissection scope in Fig. 1.2 gives you the option of using an overhead lamp, a lamp in the base, both, or neither. This knob is used to make that selection. Light intensity is not adjustable on this scope.

Rules of Microscopy

There are a few fundamental rules of microscopy everyone should follow. It will be immediately obvious to other microscopists that you know your way around a microscope when you apply these rules. The rules deal with microscope safety, care, and operation. It is important that you follow these rules, because doing so protects the microscopes, and even student grade microscopes are quite expensive.

Task #3: Learn and apply the following rules of microscopy.

1) ALWAYS use two hands whenever you carry a microscope: one hand goes under the base, and the other hand is used to grasp the arm of the scope. Cradle the microscope in front of you as you carry it.
2) Use ONLY lens paper to clean lenses. Any other kind of paper or cloth can scratch lenses on scopes. Clean lenses only when they need it.
3) ALWAYS put the lowest power objective lens of a compound scope in place when you start looking at a specimen. Also put the lowest power objective lens in place when you are finished.
4) When you prepare to focus on a slide ALWAYS rotate the lowest power objective lens into position and move the stage to its highest position. Then use the coarse focus knob to move the stage down and away from the objective lens as you locate the plane of focus. Move the stage to its lowest position and make sure you remove the slide when you are done with the scope for the day.
5) NEVER rotate the fine focus knob more than one full rotation in either direction as you attempt to focus on a specimen. If you need to rotate the fine focus knob more than one full turn in either direction go back and use the coarse focus adjustment to find the plane of focus, and then use the fine focus knob to achieve final focus.
6) ALWAYS work your way up through the objective lenses from lowest power to the desired power, in order. If you skip a lens you may have a difficult time finding the plane of focus again. If you are using the dissection scope, start with the magnification adjustment ring at 0.7x and increase magnification from there.
7) Use immersion oil ONLY with lenses designed for that purpose. Objective lenses that have a black ring around the tip of the lens can be used with immersion oil. The instructor should demonstrate how to use immersion oil before you use it.

Procedure for setting up a compound scope

1) Check the microscope
   - Make sure the lamp works
   - Move the stage to its lowest position
   - Make sure a slide was not left on the stage by the previous user
   - Rotate the lowest power objective lens into place
   - Check the lenses to make sure they are clean. Clean them only if needed.
2) Adjust the condenser
   - Focus on a slide.
   - Use the condenser adjustment knob to move the condenser to its highest position
   - Place the tip of a probe through the notch at the front of the lamp housing so it rests lightly but directly on the center of the lamp.
   - Look through the eyepiece while holding the probe tip in position and rotate the condenser adjustment knob, slowly lowering the condenser until you see a sharp outline of the silhouette of the probe tip in the field of view.
   - When the silhouette of the probe tip is in sharp focus you may also see a light mottling pattern in the rest of the field of view. This mottling is produced by imperfections in the glass surface of the lamp. Use the condenser adjustment knob to move the condenser either up or down just enough to lose the background mottling. Your condenser is now correctly adjusted.

Task #4: Make a wet-mount slide

You will need to make your own temporary (wet-mount) slides from time to time. To make one you need a specimen, a microscope slide, a cover slip, a pipette or eyedropper, and some plasticene clay.

Place your specimen in the middle of a glass slide and use a pipette to place a couple of drops of water on top of the specimen. Make sure that there are no air bubbles clinging to the specimen. Next, hold some plasticene clay in one hand and a cover slip in the other. Gently drag each corner of the cover slip across the surface of the clay. A small amount of clay should now be on each corner of the cover slip. These small bits of clay serve as a tiny pillars or support posts that keep the weight of the cover slip from flattening the specimen. Lay one edge of the cover slip down so it extends from one side of the microscope slide to the other. Support the opposite edge of the coverslip with a probe tip. Lower the coverslip slowly until it comes in contact with the drop of water covering the specimen. Capillary action will draw the water out between the coverslip and the glass slide as you lower the coverslip all the way down. Lower the cover slip slowly in order to minimize the number of air bubbles that get trapped under the coverslip.

If the cover slip is floating because there is too much water underneath it, take the corner of a paper towel and touch it to the space between an edge of the coverslip and the glass slide. The coverslip will settle onto the glass slide as water wicks out. If there is too little water under the coverslip because it does not reach all edges of the coverslip, add more water with a pipette. Water is drawn in by capillary action.

Your slide is now ready for observation. Keep in mind that the coverslip is not physically attached to the glass slide, so you need to handle the slide gently. Follow your instructor's directions about what to do with the glass slide and the coverslip when you are done.

Be professional; don't leave a mess.

Group Questions

1) What is the difference between magnification and resolving power of a microscope?
2) What happens to the apparent orientation of your specimen in the field of view when you observe it under a compound microscope? Under a dissection microscope?
3) Why use different kinds of microscopes? Describe similarities and differences between compound and dissection microscopes.

Chapter 2: Phylogenetic Analysis

Scientists use phylogenetic analysis, also known as cladistics, to test hypotheses about relatedness within and between taxa. You will almost certainly have the opportunity to hear a wide selection of talks about studies that employ this method if you attend a professional conference for zoologists. This is because taxonomists use this method to classify new species and to test the scientific validity of taxa. Many taxa have been around since the 1700s and need review.

The goal of today's exercise is to introduce you to the basics of cladistics and to give you a chance to work through an exercise where you develop a phylogenetic tree for a selected group of invertebrate taxa. Before you can do this you need to know the vocabulary of this discipline. Take time before lab to become familiar with the following list of terms.

<u>Terms used in phylogenetic analysis</u>

1. **Taxon** – any scientifically named group of organisms. A taxon can be as small as a species or as large as a domain.

2. **Ancestral taxon** – a taxon that gave rise to at least one descendant taxon by speciation.

3. **Daughter or Descendant taxon** – a taxon that descended from an ancestral taxon by speciation.

4. **Monophyletic group or Clade** – a group that includes an ancestral taxon and all of its descendant taxa. The identification of monophyletic taxa is the goal of cladistics.

5. **Paraphyletic taxon** – a violation of being monophyletic because at least one descendant species of an ancestral taxon is not included. An example of this Linnaeus' Class Reptilia. His Reptilia contained turtles, snakes, lizards, and alligators, but not birds. His Reptilia was identified as being paraphyletic because we discovered that birds and all other reptiles share a common ancestor. Because birds are now included in the revised Clade Reptilia, it is no longer paraphyletic.

6. **Polyphyletic taxon** – a violation of being monophyletic because more than one ancestral taxon is required to describe the origins of the taxa within the group. A good example of a polyphyletic taxon is the now defunct Kingdom Protista. This kingdom contained slime molds, protozoans, algae, and just about everything else that doesn't fit conveniently in any of the other kingdoms. Taxonomists are currently working to unravel this twisted mass of taxa, but for the present it still takes multiple ancestral taxa to explain the origins of the diversity in the Protista, e.g., one for slime molds, another one for algae, etc.

7. **Ingroup** – a group of taxa of interest, i.e., the taxa you want to analyze.

8. **Sister group or Sister taxon** – the most closely related taxon to the ingroup that is not part of the ingroup. For example, if your ingroup is all mammals then your sister taxon is reptiles.

9. **Outgroup** – a taxon that is included in a phylogenetic analysis for comparative purposes, usually to help identify polarity of traits. The sister taxon is ideally used as the outgroup.

10. **Character** – an anatomical or genetic trait of a taxon. Do not use ecological characters such as "lives in water." Instead use the anatomical character that led you to that conclusion, e.g., "has fins". Be advised that characters can be gained or lost during evolution, but the same character cannot be gained, lost, and then regained.

11. **Symplesiomorphy** – a character that ancestral and descendant taxa share, it's a shared ancestral trait. A symplesiomorphy of mammals is the backbone, because mammals and its ancestors including fish, amphibians, and reptiles also have a backbone. Do not use symplesiomorphic characters to do cladistics.

12. **Synapomorphy** – a character that a descendant taxon has that its ancestral taxon did not, it's a shared derived trait. An example of a synapomorphy for mammals is hair. All mammals have hair, but its ancestral taxon did not. Use only synapomorphic characters when doing cladistics.

13. **Character polarity** – an informed decision about the relationship between different versions or states of a particular character. We use the outgroup to determine which version of a character is ancestral and which is derived. An example of this is light-sensory organs. By looking at the structures used by animals to sense light we determine that the character polarity of this character goes from a light sensory patch of tissue to a cup-shaped eye to a constricted cup-shaped eye to a lens-bearing eye as being the progression from ancestral to derived character states.

The goal of cladistics is to describe the evolutionary relatedness between taxa and to solve problems of paraphyletic and polyphyletic taxa. Cladistics includes the practice of producing cladograms, also known as phylogenetic trees based on parsimony. Cladograms show evolutionary relatedness among groups included in the analysis. The most parsimonious tree is the one that has the fewest number of character changes while still preserving the existence of identifiable synapomorphies. Many different cladograms are possible if you study several taxa and several synapomorphies, but the principle of parsimony suggests that the least complex cladogram has the greatest chance of being the correct one.

Professional taxonomists do not do phylogenetic analysis by hand, instead they computer programs to generate cladograms. In this exercise, however, you will produce a cladogram by hand. This experience gives you some idea about how cladistics is done.

Phylogenetic analysis: how to construct a cladogram

The first things you need to do to carry out phylogenetic analysis are to identify the ingroup, outgroup, and synapomorphic characters to use in the analysis. Write the names of the outgroup taxon and ingroup taxa along the top of a grid matrix. Write the names of characters down the side of the grid matrix. This is the easy part.

You now need to do some research to find out which characters each taxon has and enter an appropriate notation in each cell for each taxon and character combination. This is relatively easy as long as a trait has only two options: present or not present. If a trait is present in a taxon of interest write "1" in the cell. If it is absent write "0" in the cell. Things get a bit trickier when there are multiple options for a trait. For example, these are four possible options for body symmetry: asymmetrical, radial, biradial, and bilateral. We assign these the numbers 1, 2, 3, and 4, respectively. These numbers represent the character polarity from ancestral to derived character states. Once you have filled all cells in the matrix you are ready to begin the analysis and develop your cladogram.

For the purposes of this exercise you can either choose your own ingroup and outgroup and develop a matrix for them, or you can use the provided ingroup, outgroup, and characters.

A cladogram is complete when all taxa and character combinations are accounted for on the cladogram. A cladogram is basically a branching tree rooted in an ancestral taxon. The cladogram has a long straight line called the backbone, and side branches that come off the backbone. Traits are indicated either on the backbone or on the side branches. Taxa are found only at tips of the branches.

The easiest way to explain how to do cladistic analysis is via an example.

Task #1 – Work through the following example

- Outgroup: Prokaryotes (I lumped all prokaryotes together for convenience.)

- Ingroup: All eukaryotic kingdoms – Animalia, Fungi, Plantae, Protista (Protista is no longer a viable taxon, but I use it here for the sake of convenience.)

- Characters: DNA, Nucleus, Chloroplasts (present in at least some members of the taxon), *Hox* genes, Cell walls (not peptidoglycan), and Strict multicellularity (all species must be multicellular).

The first thing you need to do when producing a cladogram is to produce a matrix of taxa in the analysis and their characters. Here are the taxa and the characters they have:

Prokaryotes: DNA

Animalia: DNA, Nucleus, *Hox* genes, Strict multicellularity

Fungi: DNA, Nucleus, Cell Wall, Strict multicellularity

Plantae: DNA, Nucleus, Chloroplasts, Cell Walls, Strict multicellularity

Protista: DNA, Nucleus, Chloroplasts, Cell Wall

Draw a matrix, i.e., a table, with a column for each taxon and a row for each character. You may do this by hand or use Excel, your choice. Write a "1" in each cell of the table where a taxon has a particular trait, and write a "0" in the cell where it does not. Once your matrix of taxa and characters is complete, you are ready to generate a cladogram.

To start, draw a long diagonal line. This is the backbone of the cladogram. A character that appears on the backbone is found in taxa farther up the tree. Any character that appears on a side branch applies only to the taxon or taxa on that side branch. Start by drawing a short hash mark across the backbone near its base and write "DNA" next to it. This indicates that all taxa on the tree have this trait, as indicated in your matrix by the complete row of 1s next to the character "DNA".

By definition, the outgroup branches off first. If you look at your matrix you see that of the characters listed "Prokaryotes" have only DNA. Draw a line extending perpendicularly to the backbone but above the hash mark for DNA and write "Prokaryotes" at its tip. The taxon Prokaryotes and all of its characters are now accounted for. Now you have to decide which taxon of the ingroup branches off first.

How can you decide which of the ingroup taxa should branch off first? When you look at your matrix you see that three of the four ingroup taxa have four characters in their columns, and Plantae has five. You have to make a judgment call, OK since this may be your first try at cladistics call it a guess, and decide whether Animalia, Fungi, or Protista should branch off next. By the way, feel free to use prior knowledge and common sense as you do this. For the purposes of this example I decided to have Protista branch off next. Protista has three traits that Prokaryotes did not: Nucleus, Chloroplasts, and Cell Walls.

Now you have to decide which of these characters should go on the backbone between Prokaryotes and Protista, and which should go on the side branch to Protista. Nucleus is easy. It should goes on the backbone because all remaining taxa also have this trait. The other two traits are tougher, because all remaining taxa do not have Chloroplasts and Cell Walls. You can handle this two different ways, as indicated on the cladograms in Figures 2.1 and 2.2. Continue adding characters and branching off taxa until all taxa and characters are accounted for on the cladogram.

The cladograms in Figures 2.1 and 2.2 were both produced using the data in your matrix. Check to make sure that all taxon and trait combinations are present on both cladograms. Which cladogram is correct? Is either correct? The first question you should ask is, "Are they equally parsimonious?" They both have eight character changes, i.e., places where characters appear or disappear, so they are equally parsimonious. Even so, the cladograms show fundamentally different evolutionary scenarios.

Take time with a partner or in a small group to walk through each cladogram while referring to your matrix. Compare and contrast the two cladograms and think about why the characters were placed where they were. Discuss these differences between the cladograms. Record your observations in your lab notebook.

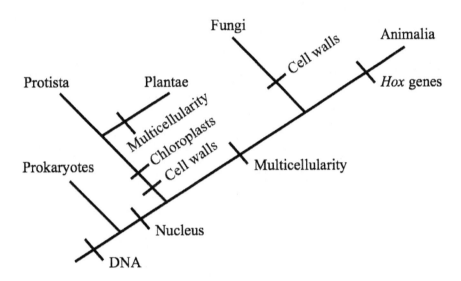

Figure 3.1. Cladogram #1, developed from the data in the matrix you produced.

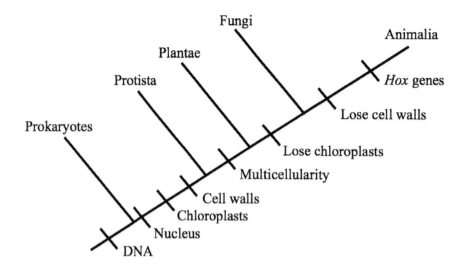

Figure 3.2. Cladogram #2, developed from the data in the matrix you produced.

Task #2

Work with one to three other people and use the data below to develop your own cladogram for the listed taxa. Don't worry if you have never done this before, most of your classmates will be in the same situation so don't hesitate to ask questions or share your thoughts and ideas.

Characters to include in the analysis

1. Cadherins (cell adhesion molecules): No=0, Yes=1.1
2. Multicellular (other than gametes and the zygote): No=0, Yes=2.1
3. Body symmetry
   3.1. Asymmetrical
   3.2. Radial
   3.3. Bilateral
4. Cnidocytes: No=0, Yes=4.1
5. Ciliated cells: No=0, Yes=5.1
6. Molted external cuticle: No=0, Yes=6.1
7. Segmented body (Note: this character evolved independently more than once): No=0, Yes=7.1
8. Deuterostomous development: No=0, Yes=8.1
9. Lophophore: No=0, Yes=9.1
10. Shell-secreting mantle: No=0, Yes=10.1
11. Water vascular system: No=0, Yes=11.1
12. Stomochord: No=0, Yes=12.1
13. Dorsal Hollow Nerve Cord: No=0, Yes=13.1
14. Chitinous hooks on unjointed legs: No=0, >4 pairs=14.1, 4 pairs=14.2

Outgroup taxon (characters from the list above are indicated)

- Choanoflagellata: 1.1

Ingroup taxa (listed alphabetically, and characters from the list above are indicated for each taxon).

- Annelida: 1.1, 2.1, 3.3, 5.1, 7.1,
- Arthropoda: 1.1, 2.1, 3.3, 6.1, 7.1
- Brachiopoda: 1.1, 2.1, 3.3, 5.1, 9.1, 10.1
- Chordata: 1.1, 2.1, 3.3, 5.1, 7.1, 8.1, 13.1
- Cnidaria: 1.1, 2.1, 3.2, 4.1, 5.1,
- Echinodermata: 1.1, 2.1, 3.3, 5.1, 8.1, 11.1
- Hemichordata: 1.1, 2.1, 3.3, 5.1, 8.1, 12.1
- Lobopodia: 1.1, 2.1, 3.3, 7.1, 14.1
- Mollusca: 1.1, 2.1, 3.3, 5.1, 10.1
- Nematoda: 1.1, 2.1, 3.3, 6.1
- Platyhelminthes: 1.1, 2.1, 3.3, 5.1
- Porifera (not a taxon, but used for convenience in this exercise): 1.1, 2.1, 3.1
- Tardigrada: 1.1, 2.1, 3.3, 6.1, 7.1, 14.2

Let's do the first few steps together. First, generate a matrix of taxa and characters from the information above if you have not already done so. Fill in each cell of the matrix, indicating which taxa have which characters. Put a "0" in a cell when a taxon lacks a character. You are ready to start generating a cladogram when the matrix is complete.

To begin draw a backbone, just like you did for the previous example. Choanoflagellata is the Outgroup, so by definition it should be the first taxon to branch off of the backbone. Look in the column in your matrix below Choanoflagellata and you will see that it has only character 1.1 (look back at the list of characters as you make your cladogram and remind yourselves which characters are which). Make a hash mark near the bottom of the backbone of the cladogram and write "1.1" next to it. Next, draw a line perpendicular to the backbone and just above that hash mark and write "Choanoflagellata" at the end of that line. You are now done with the Choanoflagellata.

Ingroup taxa are listed alphabetically, so they are almost certainly not in the order they will appear on the completed cladogram. One way to move forward from this point is to make a copy of the matrix and cut it into vertical strips with each taxon having its own strip. Arrange the strips so that the taxa with the fewest characters and the lower numbered character states are farthest to the left (closest to the Outgroup), and taxa with more characters and larger character state numbers are farthest to the right. When you do this you'll probably see that Porifera is a good candidate to line up next to Choanoflagellata. Porifera has only three characters, and it shares one of them with Choanoflagellata. The shared character is already written on the backbone of your cladogram so you don't need to write it again for Porifera. Porifera, however, also has characters 2.1 (multicellular) and 3.1 (Asymmetrical body). Make another hash mark on the backbone just above the line going to Choanoflagellata and write 2.1 next to it because Porifera and all other taxa have character 2.1, but what about 3.1, asymmetrical body? Should this character go on the backbone or on the side branch going to Porifera? Discuss this question as a group and then place character 3.1 on the cladogram where you think it works best. Draw another line perpendicular to the backbone above the hash mark for 2.1, and write "Porifera" at the end of it. You are now done with Porifera in the analysis.

Whenever you enter a higher character state for a particular character on the backbone it replaces the earlier, ancestral character state. For example if you put character 3.1, asymmetrical body, on the backbone it will be replaced by character 3.2, radial symmetry, which will later be replaced by character 3.3, bilateral symmetry. The assumption is that all taxa farther up the backbone will have whatever the last character state was, until it is replaced by a more derived character state.

Repeatedly check your cladogram as you add more taxa and more characters to see if a trait is more appropriate on the backbone or on a side branch. For example, taxonomists once thought that segmented bodies evolved only once and that all animals with segmented bodies were closely related. Recent research, however, strongly suggests that segmentation evolved multiple times. You will have to decide how to apply this information as you develop your cladogram.

Continue to add characters and taxa until all taxa are included on the cladogram, and all characters for each taxon are accounted for. Also keep in mind that it is possible for a trait to evolve and then disappear, but once this happens that same trait cannot reappear on your cladogram. Your goal is to produce the most parsimonious cladogram possible, i.e. the one with the fewest character changes (additions or losses of traits).

Chat with each other as well as with other groups as you add more taxa and characters to your cladogram. After all, science is a discipline of developing hypotheses (i.e., possible explanations) and then communicating conclusions. Your cladogram will be complete once all taxa are present and every character for every taxon is accounted for.

Another approach you could take to developing your cladogram is to rough out what you think the tree will end up looking like by placing the taxa where you think they should be in relation to each other, evolutionarily, and then add all of the characters and character states where they fit onto your cladogram. If you do this exercise this way the tree you start with represents your hypothesis, and the total number of resulting character changes it takes to make it work indicates its degree of parsimony.

You will probably need to make several changes to your tree before you are done, so go ahead and make guesses, make mistakes and enjoy successes. Be scientists!

Copy the final version of your cladogram into your lab notebook, and don't forget to include observations and questions pertaining to this exercise.

Group Questions

1. In what ways are the cladograms in Figures 2.1 and 2.2 alike? In what ways are they different?
2. Reflect on the cladograms in Figures 2.1 and 2.2, and describe the different evolutionary scenarios they present.
3. Comment on any unexpected results that appeared in your final cladogram of the invertebrates.

Chapter 3: Sponges

There are about 10,000 described species of living sponges. Sponges have been around for well over 500 million years. Sponges have anatomically simple bodies and are widely viewed as the basal group that gave rise to all other animals. Sponges lack true tissues as adults; they lack cell junction molecules, and are therefore considered to be Metazoans - animals, but not Eumetazoans – animals with true tissues.

Sponge characteristics include the following:

1) Choanocytes – flagellated collar cells
2) Cadherins – cell junction molecules, present only in larvae
3) Aquiferous system – water is pulled into the body through many tiny openings called ostia and then past choanocytes where small particles in the water are captured as the sponge carries out suspension feeding. Water then leaves the sponge after passing though a series of channels and exits via larger openings called oscula.
4) Cellular digestion – these animals lack a gut
5) Spicules – $CaCO_3$ or SiO_2, produced by sclerocytes in the mesohyle, the space between the outmost layer of cells and the layer of cells that lines channels of the aquiferous system.
6) All sponge cells are totipotent – that is, all sponge cells are stem cells, and sponges use this characteristic to constantly refine and adjust their bodies as cells around and change shapes and functions as needed. They also use this trait to carry out clonal growth by budding and fragmentation.

Though sponges are anatomically simple, ecologically sponges are amazing. While most animals deal with ecological challenges via complex behavioral responses, sponges can't do that. All they can do is sit there attached to a rock or something. Though sponges don't employ complex behaviors, they are masters of chemical warfare. Biologically active molecules isolated from sponges have been discovered that are anti-viral, anti-bacterial, anti-fungal, immunosuppressant, neuroinhibitory, and a suite of other impressive biochemical capabilities. These molecules allow sponges to fend off most pathogens, predators, and competitors.

All sponge taxa were historically lumped into the now largely obsolete Phylum Porifera, though the term Porifera is often retained for convenience when we want to refer to all sponges. Sponges are assigned to two clades depending on their cellular anatomy: Symplasma and Cellularia. Clade Symplasma includes the hexactinellid (glass) sponges. This small group of about 400 species has syncytial tissues and produces a highly complex skeleton of siliceous (SiO_2) spicules. Clade Cellularia, which includes the vast majority of extant sponges, all have cellular construction. There are two subgroups within Clade Cellularia: Demospongiae and Calcarea. Demospongiae includes the majority of cellular sponges, all of which are made up of individual cells and produce siliceous (SiO_2) spicules, while Calcarea is a smaller group that produces calcium carbonate ($CaCO_3$) spicules.

Sponges exhibit three different body plans. They are referred to as asconoid, scyonoid, and leuconoid plans. These body plans are not exclusive to any clade, so they are not useful when it comes to resolving taxonomic questions. Understanding differences between these body plans, however, provides interesting insights into how sponge bodies work.

Clade Symplasma

Euplectella is a genus in Clade Symplasma. It has a highly unusual lifestyle for a sponge. It lives in deep water where muds and oozes are the dominant substrates. Most sponges require a hard substrate for attachment and survival, but this sponge produces a highly complex endoskeleton comprised mainly of SiO_2 spicules that allows it to survive on soft substrates.

Task #1:

1) Use a magnifying lens or dissection scope to study the skeleton of *Euplectella*. If you do not have access to skeleton of *Euplectella* go online to find photographs or use the images in Figures 3.1 and 3.2 to complete this portion of the lab.
2) DRAW a small section of the spicule body wall. Be sure to include a scale bar and observations and questions to accompany your drawing. Note: This reminder will not be repeated for other drawings you are directed to include in your laboratory notebook.

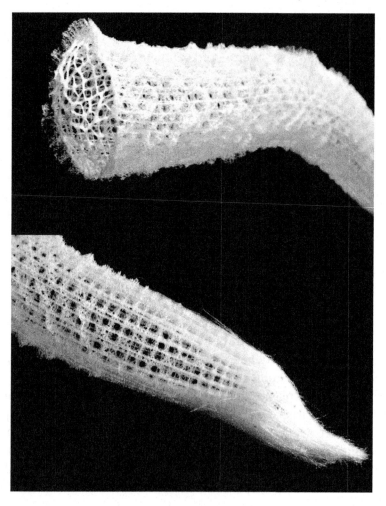

Figure 3.1. Skeleton of the glass sponge *Euplectella*: top of sponge, above; base of sponge, below (Photographs in the public domain, courtesy of the United States Geologic Survey).

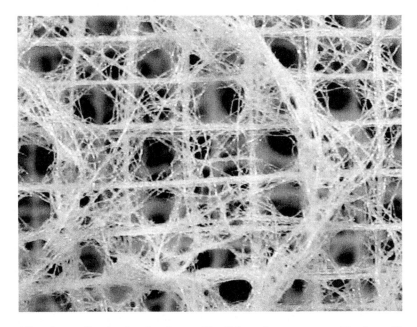

Figure 3.2. Magnification of spicules in the wall of the glass sponge *Euplectella*. (Photograph courtesy of an anonymous contributor, available under the Creative Commons Attribution Share-Alike 2.5 Generic License.)

Clade Cellularia

Asconoid plan

Leucosolenia is a good example of the asconoid body plan. The asconoid plan is the least anatomically complex of the body plans found among the cellular sponges. The asconoid plan has a thin body wall separating the external environment from the internal chamber known as the spongocoel. The body of *Leucosolenia* includes many tiny incurrent openings called ostia, an interconnected cluster of choanocyte-lined tubes, and multiple excurrent openings called oscula.

Task #2:

1) Take a cluster of *Leucosolenia* and immerse it. Observe the sponge's shape with a magnifying glass or dissection microscope. DRAW a small part of the cluster. Use Figure 3.3 to help you identify what you see. Be sure to include at least one osculum in your drawing.
2) Make a wet-mount of a small piece of a preserved specimen of *Leucosolenia*. DRAW the organization of spicules in the body wall. Use Figure 3.4 to help you identify what you see.

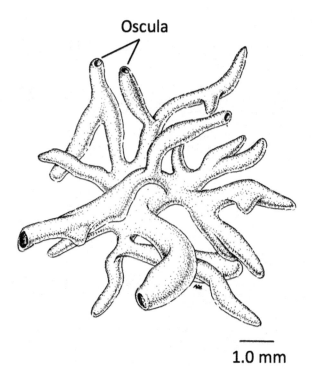

Figure 3.3. The sponge *Leucosolenia*.

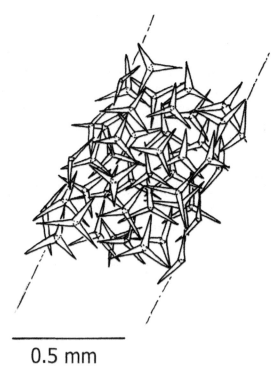

Figure 3.4. The arrangement of tri-radiate spicules in the body wall of *Leucosolenia*.

Syconoid plan

Scypha and *Grantia* are the two most commonly available genera of sponges with the syconoid body plan. The syconoid body plan is different than the asconoid plan in that the body wall is made up of a series of invaginations extending inward toward the spongocoel. These are the incurrent canals. There are also evaginations extending outward away from the spongocoel. These are lined by chaonocytes and are the feeding chambers.

Task #3:

1) Immerse a preserved individual *Scypha* or *Grantia* and observe it with a magnifying glass or dissection scope.
2) DRAW an entire sponge. Use Figure 3.5 to help you identify what you see.
3) Observe a prepared cross-section slide of the body of *Scypha* or *Grantia*.
4) Observe a prepared cross section slide of *Scypha* or *Grantia* showing eggs.
5) Observe a prepared cross-section slide of *Scypha* or *Grantia* showing embryos.
6) DRAW a cross section of the syconoid body plan with eggs in one choanocyte chamber and embryos in another. Refer to Fig. 3.6 to help you identify what you see.

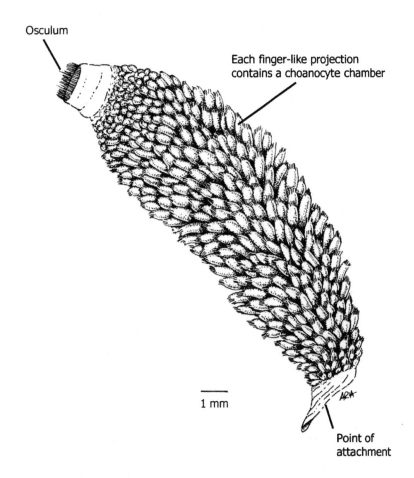

Figure 3.5. *Scypha* whole body.

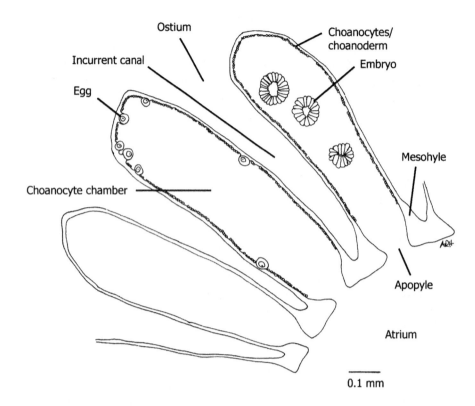

Figure 3.6. A composite drawing of a cross-section though the sponge *Scypha*, with one choanocyte chamber with eggs and another chamber with embryos. Atrium = spongocoel.

Leuconoid plan

The leuconoid plan is the most anatomically complex of the three sponge body plans. In this body plan water enters the sponge through tiny incurrent openings called ostia that open into incurrent canals. Water travels through these to incurrent canals to spherical, choanocyte-lined chambers. Water exits the choanocyte chambers via excurrent canals that fuse with each other forming progressively larger canals that empty into a spongocoel and finally exits the body through one of many oscula. The feeding chambers are so small that there can be thousands of them cm^{-3}, and are observed most easily via electron microscopy.

Bath sponges are members of the Demospongiae. These sponges have skeletons made of spicules and a stiff protein material called spongin. Spongin is what makes bath sponges "spongy." The stiffness and shape of spongin varies from species to species.

Task #4:

1) Observe a bath sponge with the naked eye, a magnifying lens, and then a dissection microscope. Describe what you see.
2) Make a wet mount slide of a tiny piece of spongin from a bath sponge. Observe it under a compound scope. DRAW some of the spongin. Use Fig. 3.6 to help you identify what you see.
3) Use a compound microscope to observe a prepared slide of gemmules from a freshwater sponge. Gemmules are overwintering bodies, not embryos or reproductive cells.

Freshwater sponges produce them when environmental conditions deteriorate. DRAW a gemmule and use Fig. 3.7 to help you identify what you see.

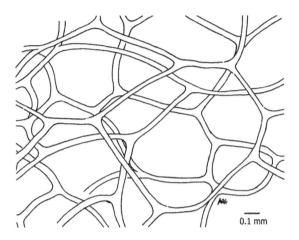

Figure 3.6. Spongin from a bath sponge.

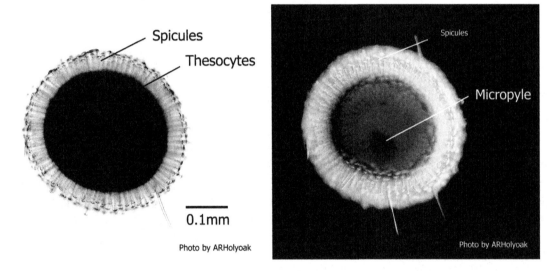

Figure 3.7. Photomicrographs of gemmules from a freshwater sponge: bright field image (left), and dark field (right).

Group Questions

The current paradigm suggests that sponges were the first animals, there were no predators when sponges evolved, and that available food was floating in the water. Keep these things in mind as you answer the following questions:

1) Develop a hypothesis that explains the origin of sponge skeletons, i.e., why did sponges develop skeletons in the first place?
2) List advantages and disadvantages of asconoid, syconoid, and leuconoid sponge body plans.

Chapter 4: Cnidaria

There are over 10,000 living species in Clade Cnidaria, including jellyfish, corals, sea anemones, and their kin. Cnidarians are recognized as basal eumetazoans. Their bodies, though anatomically simple, contain true tissues and a gut where external digestion takes place.

Characteristics of the Cnidaria include the following:

1) Cnidocytes containing a large organelle called a nematocyst – these are the stinging cells.
2) Ciliated planula larva
3) True tissues with cells connected to each other via cadherins (cell junction molecules)
4) Diploblastic body plan – two embryonic tissue layers, ectoderm and endoderm that give rise to the epidermis and gastrodermis respectively.
5) Radial symmetry of the body or modules (functional units) of the body
6) Mesoglea – an acellular gelatinous layer between the two tissue layers
7) Gastrovascular cavity – a sac-like gut that has one opening that is both mouth and anus.
8) Polyp and medusa stages (in most)

Cnidarians are carnivores. They use tentacles bearing batteries of cnidocytes to subdue and capture prey. Captured prey are stuffed into the gastrovascular cavity where more cnidocytes inject toxin into them and extracellular digestion takes place. Cnidocytes are also used for defense and competition.

The nematocyst of a cnidocytes fires a hollow tube into its prey through which toxins are injected. Toxin potency varies greatly between species. Toxins of some species such as the common west coast sea anemone *Anthopleura elegantissima* are so benign that you can run your finger along their tentacles and feel nothing more than a little stickiness. OK, you would get a numb tongue if you licked one, but that would be just plain crazy. Other species have toxins so powerful that they can kill humans. The most notorious of these is the box jelly *Chironex fleckeri*, the sea wasp.

The majority of cnidarians are marine, but a few species live in freshwater. Clade Cnidaria is divided into four smaller main taxonomic groups: Hydrozoa, Scyphozoa, Cubozoa, and Anthozoa.

Hydrozoa

Most members of this group are relatively small or their functional units, also called zooids, are small. They usually have a clonal colony-forming polyp stage, and a non-cloning gonochoric, i.e., sexually reproducing, medusa stage. Medusae, i.e., jellyfish, in this taxon are called hydromedusae. These medusae are smaller and anatomically different than "true jellyfish" of the Scyphozoa. Hydromedusae have small, tall bells, and a constricted opening at the base of the bell formed by a shelf of tissue called the velum. Other common characteristics of hydrozoans include solid tentacles and gonads that bulge outward from the body wall when they are ripe.

Hydra is a commonly studied freshwater hydrozoan even though it lacks a medusa stage and lives as physiologically independent polyps. *Hydra* does, however, carry out sexual and clonal reproduction, both of which are typical for hydrozoans.

Obelia is a commonly studied marine hydrozoan. It is much more representative of the hydrozoan body plan and life cycle than *Hydra*. *Obelia* polyps grow as clonal polymorphic colonies. The free-swimming medusa stage of *Obelia* is gonochoric.

Physalia, the Portuguese man-of-war, is a free-floating colony belonging to a group called siphonophores. Siphonophores are actually floating colonies of zooids, and the different types of zooids cooperate to meet the colony's needs.

Tasks

Hydra

1) Use a magnifying glass or dissection microscope to observe a live specimen of *Hydra*. DRAW its body form and describe its behavior.
2) Observe a prepared cross-section slide of *Hydra* and DRAW what you see. Use Figure 4.1 to help you identify what you see.
3) Observe prepared slides of budding, male, and female *Hydra*. DRAW a composite figure that shows budding, female, and male reproductive structures. Use Figure 4.2 to help you identify what you see. Note: Sexually reproductive individuals can have multiple ovaries or testes.
4) Observe the arrangement of cnidocytes on a tentacle of *Hydra*. Use Figure 4.3 to help you identify what you see.

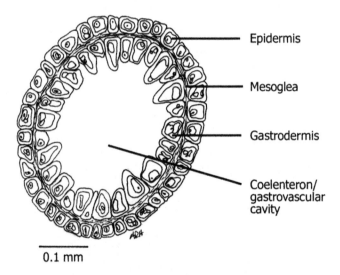

Figure 4.1. Cross-section through the gastrovascular cavity of *Hydra*.

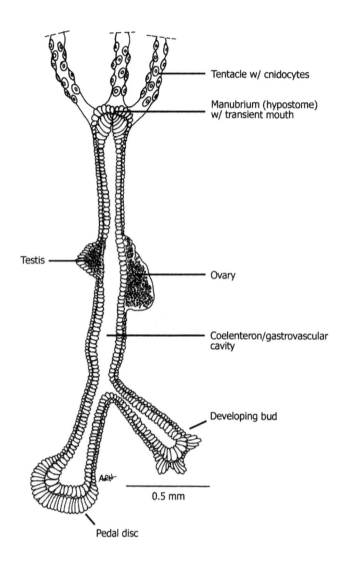

Figure 4.2. Composite drawing of a longitudinal section of *Hydra*, including a bud, an ovary and a testis.

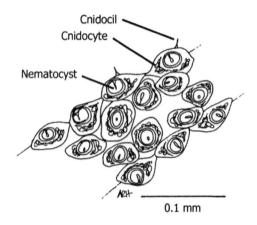

Figure 4.3. Cnidocytes on a tentacle of *Hydra*.

Obelia

1) Observe a prepared slide of an *Obelia* colony. DRAW and label a portion of the colony, include at least one gastrozooid and one gonozooid in your drawing. Use Figure 4.4 to help you identify what you see.
2) DRAW an *Obelia* medusa from a preserved specimen or prepared slide. Use Figure 4.5 to help you identify what you see.

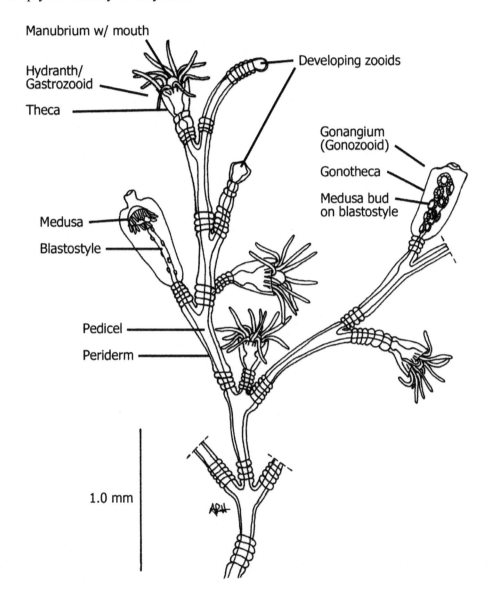

Figure 4.4. *Obelia* colony (polyp).

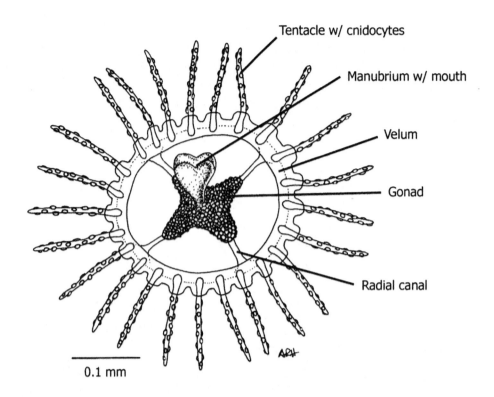

Figure 4.5. Anatomy of *Obelia* hydromedusa, subumbrellar view.

Physalia

1) Observe a preserved specimen of *Physalia*. RECORD your observations and questions about this variation on the hydrozoan body plan.

Scyphozoa

Scyphozoans include the "true jellyfish". Most members of this group spend the majority of their life cycle as gonochoric medusae, and a relatively short time as small, sessile, clonal polyps. Some members of this group, however, skip the polyp stage entirely.

Tasks

Aurelia

Aurelia is commonly called the moon jelly, and it is a good representative of the Scyphozoa. It has a small sessile polyp stage and a large medusa.

1) Immerse a preserved adult *Aurelia* medusa, observe and DRAW it. Use Figure 4.6 to help you identify what you see.

2) DRAW the following life history stages of *Aurelia* from prepared slides: planula, scyphistoma, strobila, and ephyra. Use Figure 4.7 to help you identify what you see.

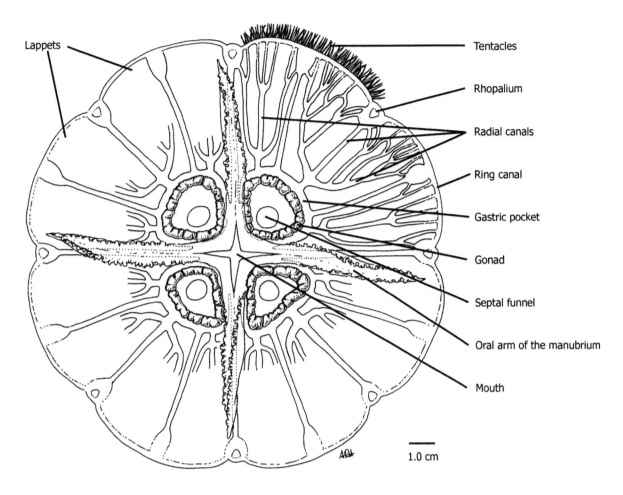

Figure 4.6. Anatomy of *Aurelia* medusa, subumbrellar view.

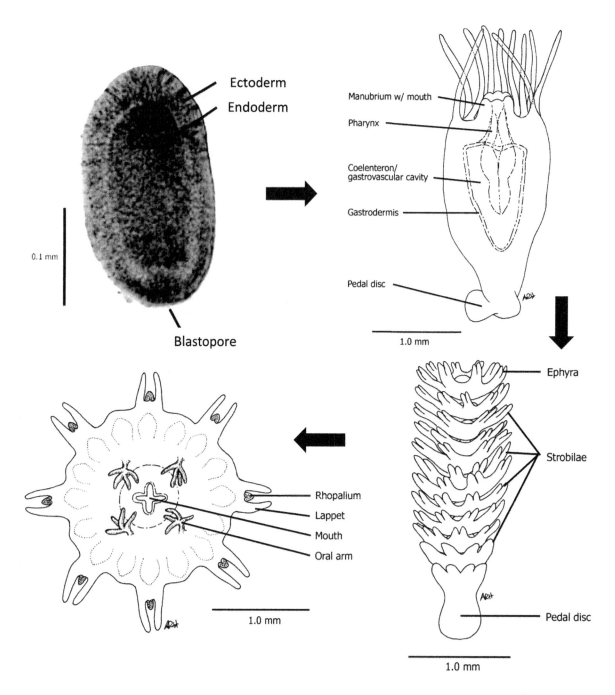

Figure 4.7. Developmental stages of *Obelia:* planula larva, top left; scyphistoma, top right; strobila, bottom right; ephyra, bottom left. Arrows indicate the order of these life stages in the life cycle.

Cubozoa

Cubozoa and Scyphozoa were lumped together taxonomically in Clade Scyphozoa until as recently as 2004, but we now consider them to be closely related but separate taxa. Cubozoans, commonly called box jellies, have a life cycle similar to scyphozoans, with a planula larva and a small benthic strobilating polyp stage. Their medusae differ significantly from those of scyphozoans. The apex of the bell of cubozoans medusae is tall and rounded, but is square in cross-section near the base. There are always four clumps of tentacles, and the opening into the bell is slightly constricted by a shelf of tissue called a velarium that is similar to the velum of hydromedusae. You are not assigned to examine a representative of this group, but you should be aware that they exist, see Figure 4.8 to see the anatomy of a cubomedusa.

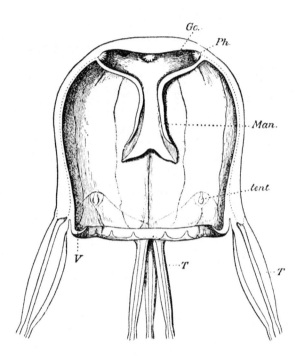

Figure 4.8. Cubomedusa of *Tripedalia cystophora*: Go, gastric ostia; Man, manubrium; Ph, nematocyst-bearing filaments; T, tentacles; tent, rhopalium; V, velarium. (Public domain image, after Conant, 1898, Mem. Johns Hopkins Univ. IV. 1)

Anthozoa

This group includes sea anemones, corals, and their relations. The anthozoan life cycle includes a gonochoric polyp that in many species also undergoes clonal growth. Anthozoan polyps are large compared to hydrozoan or scyphozoan polyps. Most anthozoans require hard substrates to survive, but a few species have adapted to life on/in soft-sediments.

Tasks

Metridium

1) Immerse an individual and DRAW its external anatomy. Use Figure 4.9 to help you identify what you see.
2) Pair up with a lab-mate and dissect two specimens of *Metridium*. Make drawings of the specimens as you carry out the dissections and expose new structures.
    a. Dissection #1: Make a longitudinal cut through the entire body column, down the length of the pharynx. Expose longitudinal septae and the gastrovascular cavity at the base of the pharynx that extends upward between complete septae. DRAW what you see. Use Figure 4.10 to help you identify what you see.

b. Dissection #2: Make transverse cuts through the body column of your specimen. Make the first cut just below the oral disc. Make a second transverse cut about 1/4 of the way down the body column so that you cut through the pharynx. Make a 3rd cut in the lower 1/3 of the body column so that you expose the gastrovascular cavity at the base of the pharynx. DRAW what you see. Use Figure 4.11 to help you identify what you see.

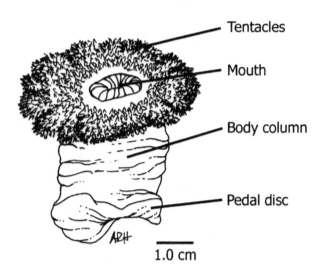

Figure 4.9. *Metridium* whole body.

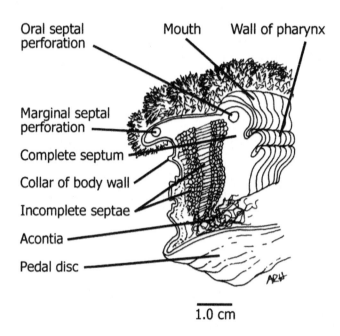

Figure 4.10. *Metridium*, one side of longitudinal section through the pharynx.

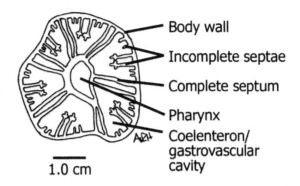

Figure 4.11. *Metridium* cross-sectional view through the pharynx.

Corals

Corals are prominent members of Anthozoa in some marine communities. You probably know them best from tropical reef systems, but they are also found in recently discovered deep, cold-water reef communities. Reef-building corals, also known as hermatypic corals build reefs. There are also soft corals that do not produce calcareous skeletons.

1) Examine the diversity of coral specimens provided. Jot down observations and questions about the diversity of coral skeletons you see.

Group Questions

The current paradigm suggests the following about Cnidarians:

· Cnidarians were the first strictly predatory animals
· They were the first animals to evolve movement via the integration of nerves and muscles

Keep these things in mind as you answer the following questions.

1) Cnidarians have been around for a long time, and they continue to be extremely successful both ecologically and evolutionarily. What attributes of cnidarians do you think allow their ongoing success? (Develop a hypothesis that addresses this question.)
2) Develop a set of hypotheses that explain when it is advantageous to a cnidarian to be colonial, and when it is advantageous to be solitary.
3) What are costs and benefits of radial symmetry?
4) As animals increase in size we tend to see more and more compartmentalization within their bodies. Do you see evidence of that trend among cnidarians? Why do you think increased compartmentalization is mandatory for increased animal body size?

Chapter 5: Platyhelminthes

There are about 30,000 described species of living Platyhelminthes, commonly called flatworms. This group includes free-living and medically important parasites.

Characteristics of the Platyhelminthes:

1) Triploblastic body, with tissues arising from embryonic ectoderm, endoderm, and mesoderm
2) Bilateral symmetry
3) Cephalization, a brain and nerve cords
4) Acoelomic body
5) Gastrovascular cavity with a single opening (in most)
6) Protonephridia or other specialized excretory cells

The flatworms are assigned to two clades, Turbellaria and Neodermata. Turbellaria are the free-living flatworms. These worms have a cellular epidermis, three layers of body wall muscles: longitudinal, circular, and diagonal, and a well-developed sensory system that includes ocelli and chemosensory cells. Turbellarians live in marine and freshwater systems, as well as in moist terrestrial habitats.

The Neodermata includes digenean flukes, monogenean flukes, and tapeworms. These are all parasites, but that's not why they are in a different clade than the Turbellaria. They are in a separate clade because they have a different anatomy. For one thing, when they become adults they lose their cellular epidermis, and it is replaced by a syncytium comprised of portions of cells that extend up through the basal lamina. These cells are called neoblasts and are mesodermal in origin. This syncytial neodermis is highly adaptive. Why? That's one of the group questions you can look forward to answering.

Turbellaria

Turbellarian flatworms live in freshwater, marine, and moist terrestrial environments. Since these worms are free-living, they have to deal with challenges like predation, competition, and other environmental factors. Keep this in mind as you study this group.

Tasks

1) Immerse a live specimen of *Planaria* or *Dugesia* and use a magnifying lens or dissection scope to observe and describe its body form and movement. Write your observations in your lab notebook.
2) Devise and carry out a simple experiment to determine whether live Turbellarians exhibit positive or negative phototaxis. Include your hypotheses, research methods, results, and conclusions in your lab notebook.
3) Examine and DRAW a prepared whole mount slide of *Planaria* stained specifically to highlight the digestive tract. Refer to Figure 5.1 to help you identify what you see.
4) Examine and DRAW a prepared cross-section slide through the pharynx of *Planaria*. Use Figure 5.2 to help you identify what you see.

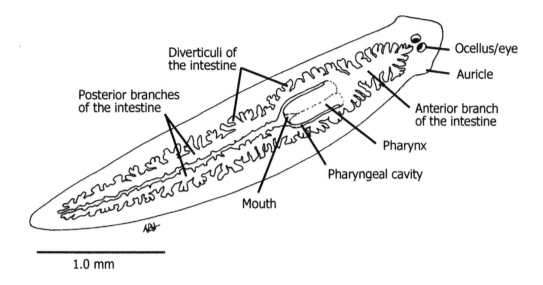

Figure 5.1. Digestive tract of the Turbellarian *Planaria*.

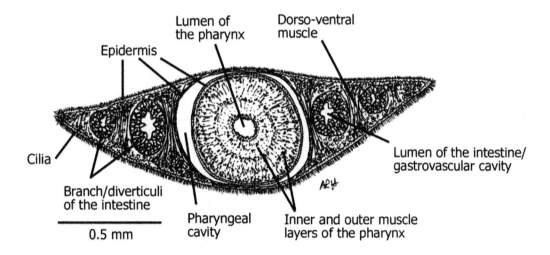

Figure 5.2. Cross-section through the pharynx of the Turbellarian *Planaria*.

Neodermata

All members of this group are parasitic, and they go through a developmental process where they replace some or all of the original cellular epidermis with a syncytial outer covering called the neodermis or tegument. Neodermata includes the clades Monogenea, Digenea, and Cestoda.

Monogenea

Monogenea are flukes that are ectoparasites. The name Monogenea means "one birth or origin". It has this name because these animals parasitize only one host during their life cycle. That is, the adult produces an immature swimming stage that seeks out its host. Once it finds a host it lives there the rest of its life.

Monogeneans parasitize fishes, amphibians, and reptiles, and they are usually species specific. These flukes attach themselves to fins, gills, and other thin tissues through which they can obtain blood and other nutrients from their host. Monogeneans are tiny and under most conditions are present in small enough numbers that they have little effect on their hosts. However, when water temperatures increase Monogenean reproductive rates increase, and they can put significant stress on their hosts as they become more heavily infested with these parasites. Monogeneans can also cause problems in aquaria or small ponds where there is limited turnover of water.

Tasks

Anesthetize a goldfish and check it for monogenean ectoparasites (anesthetization procedure after Hunt, modified from a lab developed by Dr. Edward E. Brandt, Shenandoah University).

1) Make some stock anesthetic - MS222 stock solution (0.4%, 100ml):
    - 400 mg MS-222 (Tricane)
    - 97.9 ml distilled water
    - 2.1 ml 1 M Tris-Cl (pH 9)
    - Add 5 ml MS-222 stock solution to 95 ml of clean tank water
2) Net a fish and transfer it into 100 ml of the anesthetizing solution.
3) When the fish becomes motionless remove it from the solution and do a dip rinse in distilled water to prevent premature death of the fish.
4) Place the fish in a Petri dish or small glass bowl with just enough water to cover it. Be sure that the tail is near the center of the dish.
5) Flare the tail or fin completely with two toothpicks or probe tips and keep the tail spread during examination.
6) Use a dissection scope to search for monogeneans. Begin with low magnification and focus on the surface of the tail or fins. Move to a compound microscope if needed.
7) Watch for movement by small, nearly transparent animals attached to the outer surface of the tail or fin as you zoom in. Thee small active animals are monogenean flatworms. DRAW a monogenean. Use Figure 5.3 to help you identify what you see.
8) Remember that this is a treasure hunt, and there may or may not be monogeneans present.

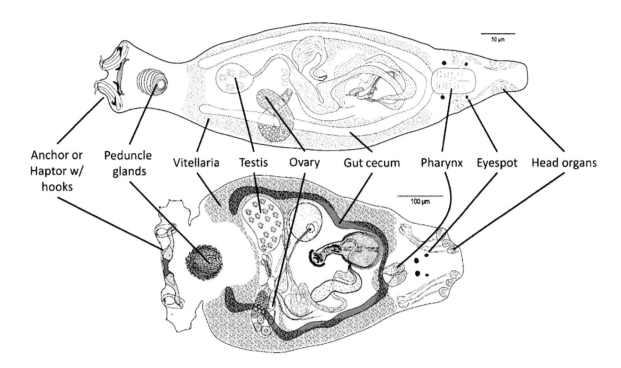

Figure 5.3. Representative monogeneans: *Calydiscoides euzeti,* above, and *Echinoplectanum leave*, below. (Images courtesy of Jean-Lou Justine, under the Creative Commons Attribution-Share Alike 3.0 Unported License and the GNU Free Documentation License ver. 1.2.)

Digenea

This group of flukes has complex life histories, i.e., they require more than one host to complete their life cycles. The name digenea means "two births or origins". Intermediate hosts are parasitized by clonal life stages, and the definitive or final host is parasitized by sexually reproductive adults.

This group is medically important because many kinds of digeneans parasitize humans. One of the most common digenean parasites of humans is *Opisthorchis (Clonorchis) sinensis*, the Chinese liver fluke. *Opisthorchis sinensis* is particularly common in areas of the Orient where the main crop is rice, human waste as used as fertilizer, and farming is done mainly by manual labor. It would be instructive to walk through this fluke's life cycle before you take time to observe its various life stages.

The Chinese liver fluke life cycle is shown in Figure 5.4. The adult lives in the liver, gall bladder, and bile duct. Adults produce huge numbers of embryonated eggs that pass through the bile duct, into the small intestine, and then out of the host body with its feces. Where human feces are used as fertilizer, the feces containing the eggs are dumped into flooded rice paddies. When the feces break down eggs float to the bottom where they are eaten by aquatic snails. Once inside a snail, the embryo is freed from its shell and it emerges as a miricidium larva. This larva changes into the sporocyst stage and produces many rediae via cloning. The sporocyst body wall breaks and rediae are released. Rediae then produce many cercariae by cloning. The cercaria stage burrows out of the snail, and swims in the water where it comes into contact with fish. The cercariae then encyst in the skin or skeletal muscle of fish as metacercariae. When a predator, including humans, eats raw or improperly cooked fish the metacercaria stage emerges, and this is where things get really interesting. The infective stage burrows out of the gut and into the abdominal cavity of the final host. It then crawls along the inner wall of the abdominal cavity in a random direction, eventually coming in contact with the liver where it burrows in, and becomes an adult.

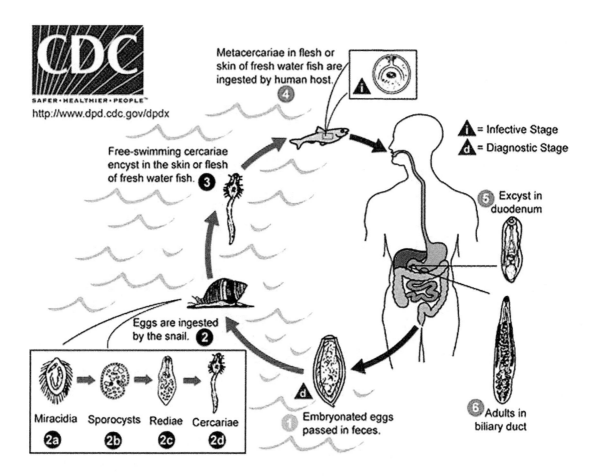

Figure 5.4. Life cycle of the Chinese liver fluke, *Opisthorchis sinensis*. (Public domain image courtesy of the Centers for Disease Control, Atlanta, Georgia, USA).

Tasks

1) DRAW as many of the life stages of *Opisthorchis* from prepared slides as are provided: egg, redia, cercaria, and metacercaria. Use Figure 5.5 to help you identify what you see. Also, locate each of these stages in the life cycle shown in Figure 5.4.
2) DRAW an adult *Opisthorchis sinensis* from a prepared slide. Use Figure 5.6 to help you identify what you see.

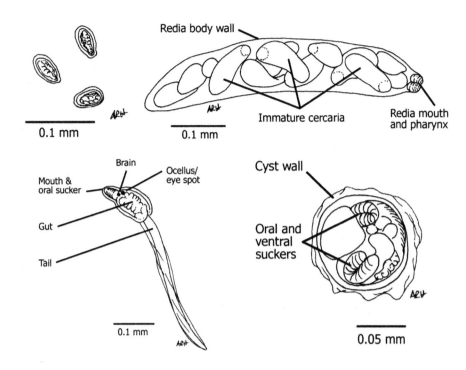

Figure 5.5. *Opisthorchis sinensis* life stages: embryonated eggs, top left; redia containing immature cercaria, top right; cercaria, below left; metacercaria, below right.

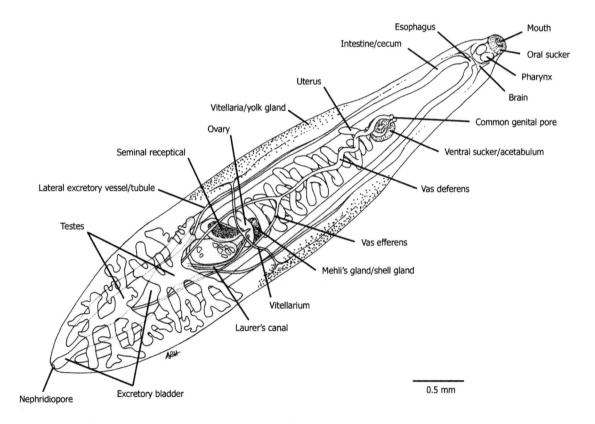

Figure 5.6. *Opisthorchis sinensis*, adult life stage.

Cestoda

Cestoda are tapeworms. Tapeworms parasitize all species of vertebrates. The longest tapeworm is found in whales and can be up to 30 m long. Some tapeworms found in humans can be up to 20 m long! Adult tapeworm bodies are so derived that they lack a mouth and digestive tract, and the adult body is dedicated almost entirely to reproduction, while the outer surface of the body is adapted to absorb nutrients directly from its surroundings – the lumen of the small intestine. It's like these worms are functionally inside out, their version of an intestinal lining is on the outside of their bodies.

Cestoda, like digeneans, have complex life histories. Their life history, however, is driven by the link between herbivores and the carnivores or scavengers that feed on them. The adult stage lives in the intestine of its definitive host and produces vast numbers of embryonated eggs in environmentally protective shells that are released with the host feces. Some of these eggs are incidentally ingested by herbivores. Eggs hatch in the gut of the herbivore and the infective oncosphere stage burrows into skeletal muscle where it encysts and becomes the cysticercus stage. It then waits. It waits for its host to die and be scavenged or to be killed and eaten. If the flesh of its host is eaten the cysticercus stage is freed from its cyst when the muscle surrounding it is digested away in the gut of its new host. It then metamorphoses into an adult, attaches itself to the wall of small intestine, and completes the life cycle.

The life cycle of the beef tapeworm *Taenia saginata,* a species found in humans, is shown in Figure 5.7. This cycle can also be completed by other species that eat beef, including dogs, wolves, rats, etc.

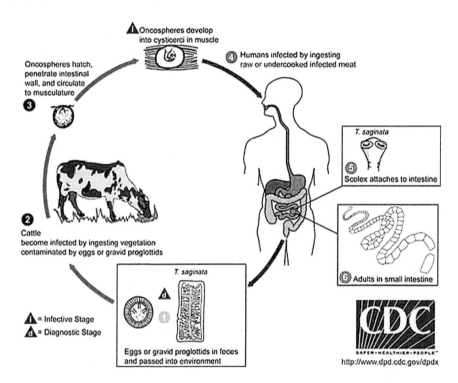

Figure 5.7. Life cycle of the beef tapeworm, *Taenia saginata*, a species found in humans. (Public domain image courtesy of the Centers for Disease Control, Atlanta, Georgia, USA)

Tasks

1) Observe prepared slides of tapeworm life stages. DRAW what you see.
2) Observe prepared slides of the scolex, a mature proglottid, and a gravid proglottid. DRAW them and use Figures 5.8 to 5.10 to help you identify what you see.

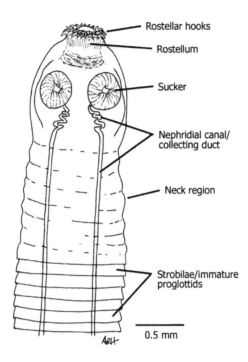

Figure 5.8. The scolex, neck, and immature proglottids of *Taenia saginata*.

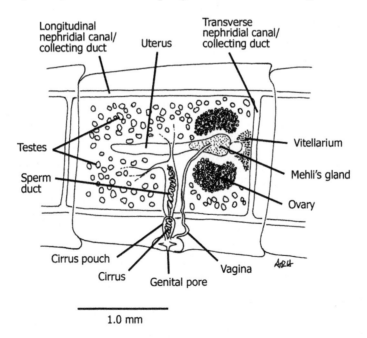

Figure 5.9. Mature proglottid of *Taenia saginata*.

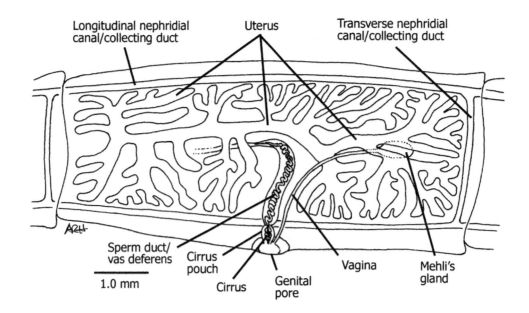

Figure 4.10. Gravid proglottid of *Taenia saginata*.

Group Questions

1) List advantages that a bilateral body plan conveys that a radial body plan does not.
2) Which anatomical characteristics do you think are advantageous to flatworms that evolved endoparasitic life styles?
3) Why is it evolutionarily advantageous for most parasites to be monoecious? Why aren't they all monoecious?
4) Explain why even relatively large flatworms (from centimeters to meters long) can do just fine without the complex circulatory systems seen in other large animals.
5) How is the neodermis a beneficial adaptation for an endoparasitic lifestyle?

Chapter 6: Mollusca

There are about 100,000 described living species of molluscs. Only Arthropoda has more, OK Arthropoda has a lot more, over one million species. Clade Mollusca includes a diverse range of variations on a common body plan. There are tiny slow creepers to large fast swimmers, and they make a living in every imaginable way. There are predators, herbivores, scavengers, filter-feeders, parasites, etc., and a few even form a temporary relationships with chloroplasts from algae that they eat. These move chloroplasts from their gut to their dorsal body wall where chloroplasts continue to carry out photosynthesis for a period of time.

Most molluscs are marine, but gastropods and bivalves also live in freshwater, and some gastropods live on land. Regardless of where they live, molluscs share the same set of anatomical traits.

Characteristics of molluscs:

1) Triploblastic body
2) True coelom
3) Mantle
4) Muscular foot
5) Visceral mass
6) Mantle cavity
7) Shell (in most)
8) Radula (in many)
9) Trochophore larva (in many)
10) Metanepridia
11) Open circulatory system (except in cephalopods)
12) Hemocyanin (copper-based blood pigment)

Clade Mollusca includes seven taxa: 1) Aplacophora – solenogasters, worm-like molluscs that lack a shell; 2) Polyplacophora – chitons; 3) Monoplacophora – deep sea molluscs with cap-shaped shells, no common name; 4) Gastropoda – snails and slugs; 5) Cephalopoda – octopus, squid, and kin; 6) Bivalvia – clams and mussels; and 7) Scaphopoda – tusk shells.

Exercises on the Polyplacophora, Gastropoda, Bivalvia, and Cephalopoda are covered in this chapter.

There is enough material in this chapter for at least two lab meetings.

Polyplacophora

There are over 900 species of Polyplacophorans, commonly called chitons. Most live in the intertidal zone or in shallow subtidal habitats, but a few species have been found in waters up to 6000 m deep. Chitons creep slowly as they use a radula to scrape bacterial and algal films off of hard substrates or sweep soft sediments into their mouths. Interestingly, there is even a predatory species, *Placiphorella velata.*

Chitons range in size from less than 1 cm to over 30 cm in length. The dorsal surface of chitons is protected by eight overlapping shell plates and a leathery girdle into which the shell plates are at least partially embedded. Chitons also have a large, broad foot, and a reduced head.

Tasks

1) Examine the external anatomy of your specimen. DRAW the ventral surface of your specimen. Use Figure 6.1 to help you identify what you see.
2) Work in pairs to complete the following dissection.
    a) Place your chiton ventral surface up. Use a scalpel to make a shallow cut along the midline of the foot. Do not damage any underlying structures. Note the muscular texture of the foot. Next make a series of lateral cuts with either a scalpel or scissors and remove the entire foot, the gills and mantle cavity, thus exposing the contents of the body cavity. Rinse and then immerse your specimen in enough water to cover it. DRAW what you see in the body cavity. Refer to Figure 6.2 to help you identify what you see.
    b) Remove the digestive gland. It is the material between the loops of the intestine. Take your time removing this organ, because the walls of the intestine are easily damaged. DRAW the digestive tract after the digestive gland has been removed. Refer to Figure 6.3 to help you identify what you see.
    c) Carefully uncoil the intestine to see how long it is, as well as how much of the body cavity is devoted to the digestive system. Next, cut around the edges of the anterior-most portion of the stomach and remove the entire digestive tract. If you are careful you should see the radular sac extending posteriorly from a mass of tissue at the anterior end of the body cavity. You should also be able to see the heart at the posterior end of the body cavity. Observe the large gonad which lies along the dorsal wall of the body cavity. Nephridia lie along the lateral walls of the body cavity. You should also a pair of branching salivary glands lying along the body wall dorsal and lateral to the radular sac. DRAW what you see. Refer to Figure 6.4 to help you identify what you see.
    d) Remove the gonad and observe the inner surface of shell plates, dorsal aorta, and other structures.
    e) Dissect the head. Use a scalpel to remove the head and associated tissues. Make a cut along the ventral midline of the head so you can lay it open and observe mirror images of the structures of the head. You should be able to see the buccal cavity, radula, radular sac, and odontophore. DRAW what you see. Refer to Figure 6.5 to help you identify what you see.
    f) Remove the radula and examine it with a magnifying class or dissection scope. DRAW what you see.

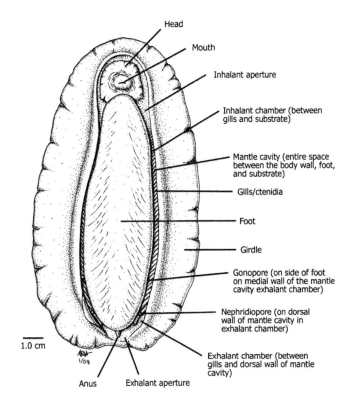

Figure 6.1. Ventral view of *Cryptochiton stelleri*.

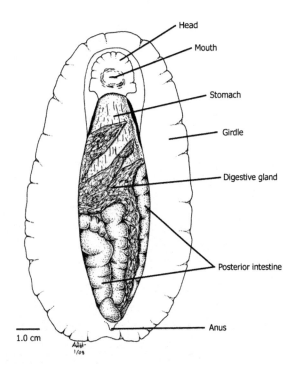

Figure 6.2. Ventral view *Cryptochiton* with the foot and mantle cavity removed, exposing the contents of the body cavity.

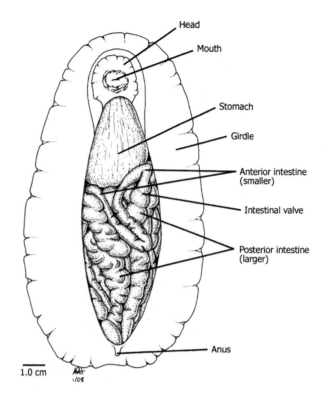

Figure 6.3. Ventral view of *Cryptochiton* with the digestive gland removed.

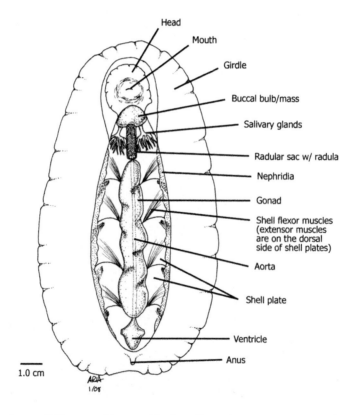

Figure 6.4. Ventral view of *Cryptochiton* with the digestive tract removed.

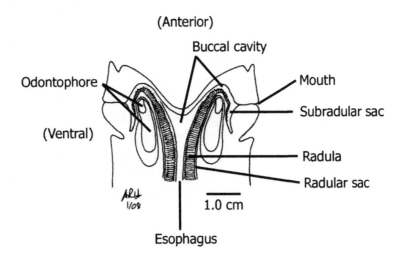

Figure 6.5. Buccal cavity and radula of *Cryptochiton*.

Gastropoda

Gastropoda is the largest group of molluscs with over 60,000 extant species. These are the snails and slugs. In this exercise you will study the terrestrial snail *Helix*. *Helix* is a well-studied snail species for at least two reasons: 1) they are easy to find within their natural geographic range and are available from most biological supply companies; and 2) they are economically important because *Helix* is used to make the French delicacy, escargot.

Tasks

1) Observe the external anatomy of *Helix*. Note the thickness of the shell and structures of the head, foot, and collar. Compare the shell of *Helix* to those of marine snails, if available. DRAW what you see. Refer to Figure 6.6 to help you identify what you see.
2) Expose the soft tissue covered by the shell. Use forceps to carefully peel the shell away from the soft tissues of the body. Do not damage the soft tissues. Take your time as you do this, because if you don't your snail could easily and quickly turn into a mangled mess. Once the shell is removed, get a wax-bottom dissection tray and fill it with enough water to cover your specimen. Use insect pins to pin the posterior portion of the foot and the margin of the anterior part of the foot to the tray. Observe the structures that were exposed when the shell was removed. DRAW what you see. Refer to Figure 6.7 to help you identify what you see.
3) Open the mantle cavity. Insert the point of a pair of scissors into the pneumostome and cut the tissue connecting the collar to the body of the snail. Continue your cut until you have separated the collar and the left margin of the mantle from the body wall. Carefully fold the dorsal wall of the mantle cavity over so you can see the structures of the inner surface of the dorsal wall of the mantle cavity, as well as structures visible through the floor of the mantle cavity. Since *Helix* is terrestrial you will see a network of blood vessels in the wall of the mantle cavity instead of gills. The wall of the mantle cavity is the site of gas exchange. DRAW what you see. Refer to Figure 6.8 to help you identify what you see.

4) Open the body cavity. Make a longitudinal cut through the dorsal wall of the head/foot from just anterior of the mantle cavity to the head. Use two pairs of forceps or a pair of forceps and a pair of scissors (<u>not</u> a scalpel) to make the cut. Pull the tissues of the body wall away from each other and, using insect pins, anchor the body wall to the floor of the dissection tray. Examine exposed structures of the body cavity before teasing anything apart. These structures comprise mainly the digestive system on the left side of the body and the reproductive system on the right side of the body. There are also many muscles, mostly along the floor of the body cavity, but these are not identified in this lab. DRAW what you see. Refer to Figure 6.9 to help you identify what you see.
5) Remove the oviduct and observe the anatomy of the posterior portion of the body cavity and of the visceral mass. Be careful not to tear or rip anything. DRAW what you see and refer to Figure 6.10 to help you identify what you see.

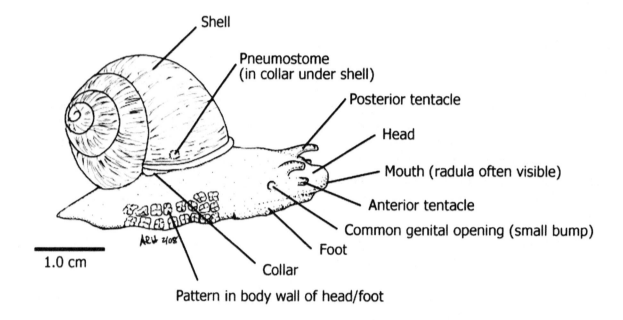

Figure 6.6. External anatomy of *Helix*.

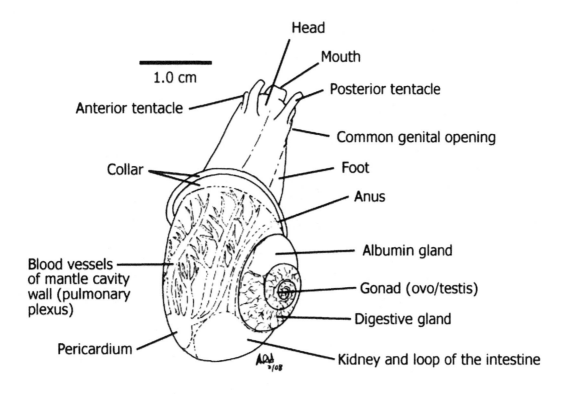

Figure 6.7. Dorsal view of *Helix* with the shell removed.

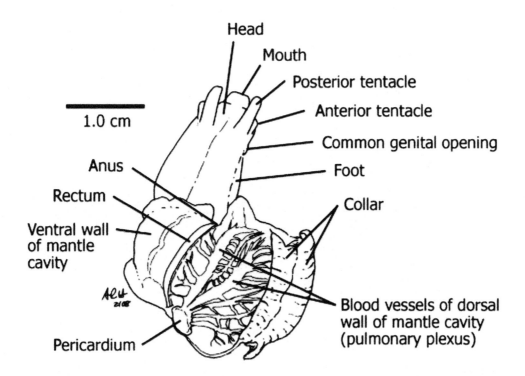

Figure 6.8. Dorsal view of *Helix* with the dorsal wall of the mantle cavity folded over to the right.

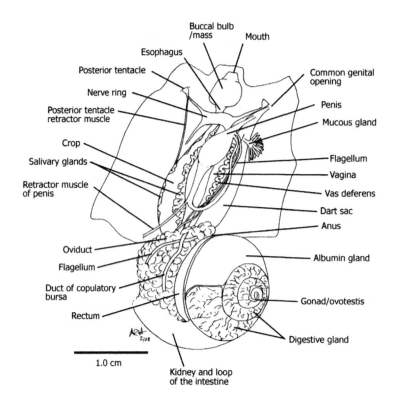

Figure 6.9. Internal anatomy of *Helix*.

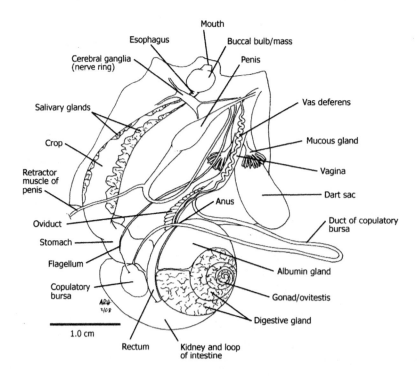

Figure 6.10. Internal anatomy of *Helix* with the oviduct removed, the digestive system pulled to the left, and the reproductive system pulled to the right. Muscles and other structures of the floor of the body cavity are not included in this figure.

Cephalopoda

There are about 800 living species of cephalopods. They are all active sight predators, and include octopi, squids, cuttlefish, and their relations. This is also an ecologically and economically important group that supports a worldwide fishery. In addition, cephalopods are almost certainly the most intelligent invertebrates. They exhibit learning, memory, and problem-solving capabilities by trial and error and by observing other individuals solve problems. Squid are pelagic, and octopi and cuttlefish are demersal.

The cephalopod body plan is highly derived compared to the ancestral mollusc plan that probably looked much like monoplacophorans. Most cephalopods have either greatly reduced shells or have lost them all together. Their foot is modified into tentacles that bear suckers or hooks, and are used for grasping prey.

Tasks

1) Examine the external anatomy of a squid. Identify the morphological and functional polarities of its body. All instructions and figures in this lab refer to the squid's <u>functional polarity</u>. DRAW what you see. Use the drawings in Figure 6.11 to help you identify what you see.
2) Examine the anatomy of the visceral mass. Place the squid on its back. Use a scalpel to make a longitudinal incision through the ventral wall of the mantle. Make your cut off-center of the ventral midline. Start the incision at the anterior ventral lip of the mantle and cut all the way to the posterior end of the mantle. Make the incision as shallow as possible while cutting all the way through the mantle wall. Open the mantle cavity carefully, little by little, and about 1/2 to 2/3 of the distance back from the anterior edge of the mantle you should see a blood vessel between the visceral mass and the ventral mantle wall. This is the medial mantle artery. You should also be able to see the medial septum which starts at the medial mantle artery and continues to the posterior end of the mantle cavity. Both the medial mantle artery and medial septum are delicate structures and will separate from their points of attachment when you open the mantle cavity. Open the mantle cavity after you have identified the medial mantle artery and medial septum. Use heavy dissection pins to anchor the walls of the mantle to the bottom of your dissection tray or cut the walls of the mantle away to keep them from flipping closed. DRAW the anatomy of the mantle cavity, and refer to Figure 6.12 to help you identify what you see.
3) Determine the gender of your squid. If your squid is male you should see a cylindrical penis attached to the left side of the visceral mass between the left gill and the leading edge of the mantle (remember your squid is on its back). You should also see a coiled spermatophoric gland just posterior to the left gill. If your squid is female you should see the funnel-shaped opening of the oviduct on the left side of the visceral mass, just anterior to the left gill. There will also be a pair of large nidamental glands located between the branchial hearts and extending posteriorly from there. Refer to Figures 6.13 and 6.14 to help you determine your squid's gender. Note: Unfortunately, sometimes it is difficult to sex squid, so check with other lab groups to see what they can see in their squid if your squid's gender is not clear.
4) Male reproductive system. To see the entire male reproductive system carefully remove the following structures: the left gill, left branchial heart, posterior vena cava, and lateral

mantle artery. Next remove the thin, transparent epithelial covering of the posterior portion of the visceral mass. DRAW the male reproductive system, and refer to Figure 6.13 to help you identify what you see.

5) Female reproductive system. The female reproductive system is largely visible when you open the mantle cavity. Remove the left nidamental gland in order to see the accessory nidamental gland under it, as well as to more clearly see the oviductal gland and oviduct. DRAW the female reproductive system, and refer to Figure 6.14 to help you identify what you see.

6) If at all possible, observe both male and female squid, perhaps swapping with another group to do so.

7) Circulatory system. Much of the circulatory system is visible when you first open the mantle cavity and expose the visceral mass, especially if your squid has been injected with colored latex. In order to see the systemic heart and circulatory structures of the gills you need to carefully remove the kidneys. Examine the circulatory system and DRAW a sketch that indicates the overall flow of blood through the body. This is particularly interesting since squid have 3 hearts! Cool, eh?

8) Digestive system. Complete your investigations of all other anatomy of the mantle cavity before doing this dissection. Remove the following structures: both gills, branchial and systemic hearts, the entire reproductive system, the ink sac (Note: be careful as you remove the ink sac, if it ruptures ink will get all over your squid – set the ink sac aside in a small bowl for later), the ventral wall of the siphon, and the siphon retractor muscles. Also, separate the head retractor muscles from the anterior portion of the digestive gland. Be advised that the esophagus is long, delicate, and easily broken. DRAW the digestive tract and refer to Figure 6.15 to help you identify what you see.

9) Observe the anatomy of the head, foot, and buccal bulb. Examine the anatomy of arms and tentacles and suckers. Also look at the anatomy of the head and mouth. Once you have done this make lateral cuts ventral to the eyes and deep enough to reach the buccal bulb. The buccal bulb is a bulbous mass of muscle and connective tissue that houses the beak, radula, mouth, and opening to the esophagus. Locate the esophagus coming out of the posterior end of the buccal bulb and then remove the buccal bulb from the head. Keep track of the dorso-ventral orientation of the buccal bulb. Make a longitudinal cut completely through the midline of the buccal bulb and lay it open showing mirror halves. DRAW what you see. Refer to Figure 6.16 to help you identify what you see.

10) If you have extra time look at the anatomy of the nervous system. Make a longitudinal cut along the midline, through the dorsal head. This will expose the dorsal portion of the brain. Refer to your textbook for help identifying what parts of the brain and cranium. You heard right...a cartilaginous cranium. Next, carefully dissect out one of the eyes from your squid. Also look for the stellate ganglia on the dorsal wall of the mantle on either side of the digestive gland (see Figure 6.12).

11) Cut the ink sac open and add a little water to the small glass bowl. Take either a crow quill pen or the squid's own pen and use it to write something in squid ink in your laboratory notebook. It is a sort of rite of passage for budding invertebrate zoologists to write with squid ink, plus it's fun.

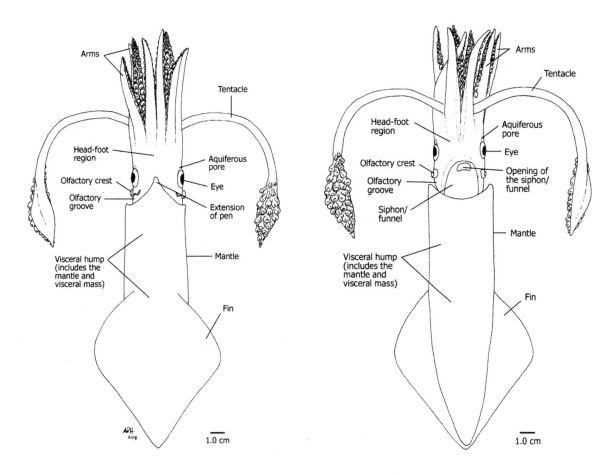

Figure 6.11. External anatomy of squid: Dorsal view, left; Ventral view, right.

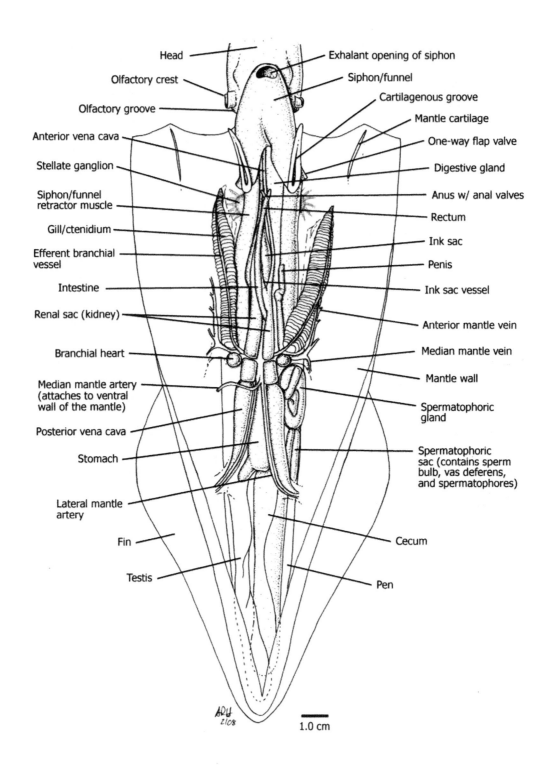

Figure 6.12. Structures of the visceral mass of a male squid.

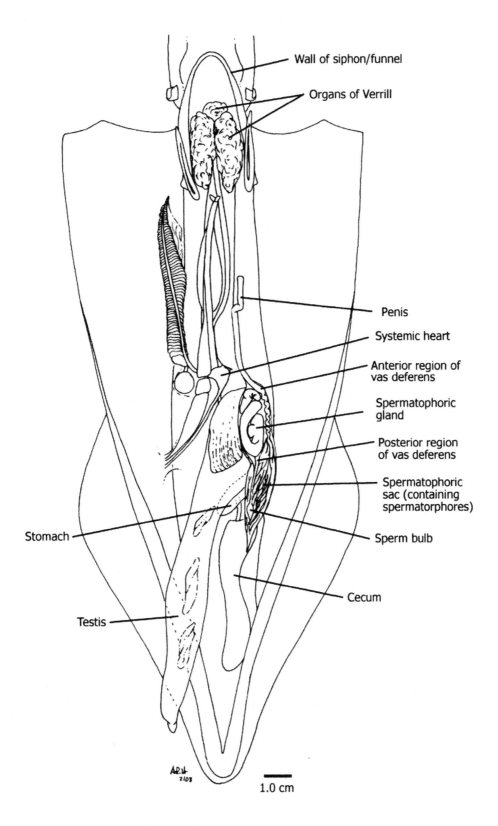

Figure 6.13. Squid male reproductive system.

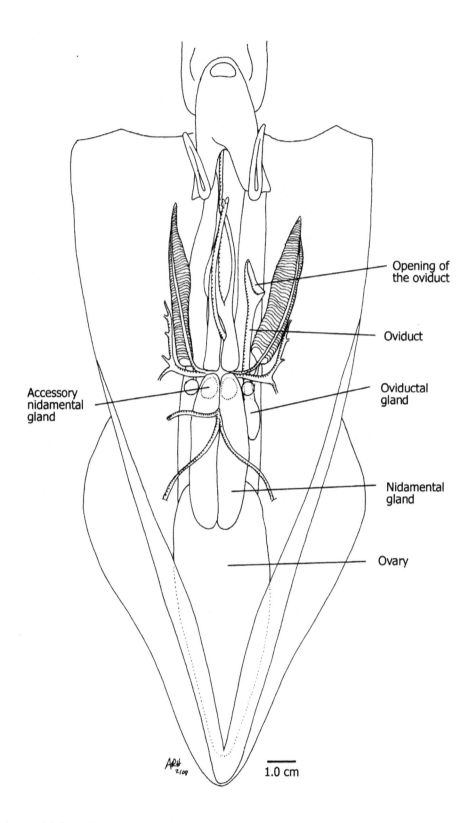

Figure 6.14. Squid female reproductive system.

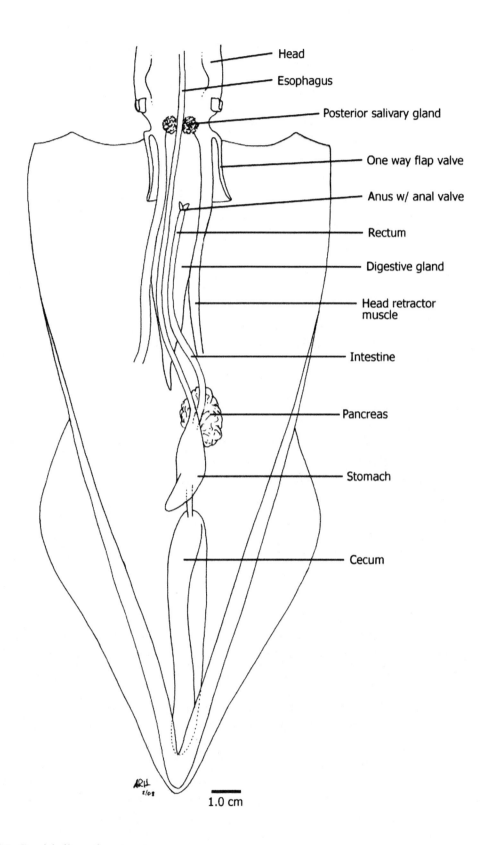

Figure 6.15. Squid digestive tract.

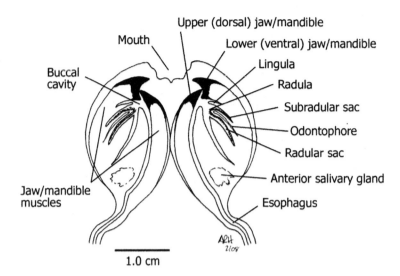

Figure 6.16. Detail of the squid buccal bulb. The ventral surfaces are toward the right and left edges of the drawing, and the dorsal surface is in the center of the drawing.

Bivalvia

The bivalves also have a highly derived body, and there are about 10,000 extant species of clams, mussels, and their relations. Most bivalves are highly specialized filter feeders that produce heavy protective shells. They also have large gills that they use for gas exchange and in most species for filter feeding. Bivalves have highly reduced heads, evident only by the location of the mouth. Bivalves live everywhere from wave-swept beaches and rocky intertidal communities to the deep sea. Some clams are adapted to live in hydrothermal vent and methane seep communities. The internal and external anatomy of bivalves shows some variability, but once you learn their basic anatomy, it is not too difficult to decipher the anatomy of most bivalves. The freshwater clam *Unio*, is the focus of this exercise, though you can use just about any marine or freshwater clam for this lab.

Tasks

1) Examine the external anatomy of the shell. Pay particular attention to the shell hinge before and after you open your specimen. The siphonal region marks the posterior end of the animal. DRAW what you see, and refer to Figure 6.17 to help you identify what you see.
2) Open your specimen and examine the anatomy of inner surface of the left shell. The best way to open the clam is to work the blade of a scalpel or a knife between the valves of the shell and cut through the anterior and posterior adductor muscles. Look at Figure 6.18 to see the location of these muscles. You will know when they have been cut because you will be able to separate the valves of the shell with relative ease. Before you pull the shell valves apart, however, use the tip of a probe to separate scrape the inner surface of the shell, thus dislodging the mantle and muscles from their points of attachment. Open the clam so that all soft tissues rest in the right valve. DRAW the inner surface of the left valve. Refer to Figure 6.18 to help you identify what you see. Again, be sure to take time to examine the internal and external portions of the hinge.

3) Immerse your clam and observe the exposed soft-tissue anatomy that is revealed when you removed the left valve of the shell. DRAW what you see. Refer to Figure 6.19 to help you identify what you see.
4) Fold the left lobe of the mantle toward the umbo and cut it away. Do not to cut into the pericardium or nephridial sac. DRAW what you see, and refer to Figure 6.20 to help you identify what you see.
5) Remove the left ctenidium (gill) and carefully shave away layers of tissue and muscles on the left side of the visceral mass, just dorsal to the foot. As you so this you will expose dull yellow and olive green masses of tissue with a small, coiled tube running through it that will be evidenced by small holes in the digestive gland and ovary. These tissues are the digestive gland (green) and gonads (yellow). The holes and spaces you hopefully see are the stomach and the intestine that winds its way through this part of the visceral mass. Find the mouth. It is located on the anterior midline of the body directly between the left and right sets of labial palps, and follow the esophagus into the visceral mass. Next, open the pericardium and nephridial sac. DRAW what you see, and refer to Figure 6.21 to help you identify what you see.
6) Examine a prepared slide of glochidia larvae of freshwater bivalves. DRAW one glochidia larva and refer to Figure 6.23 to help you identify what you see.

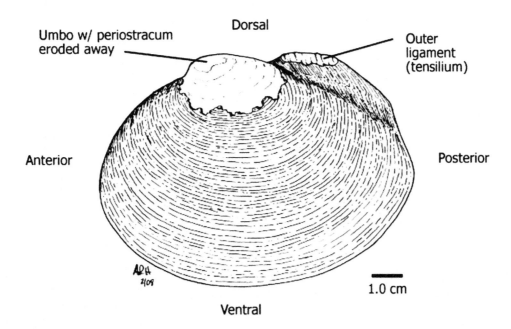

Figure 6.17. External anatomy of the left shell of *Unio*.

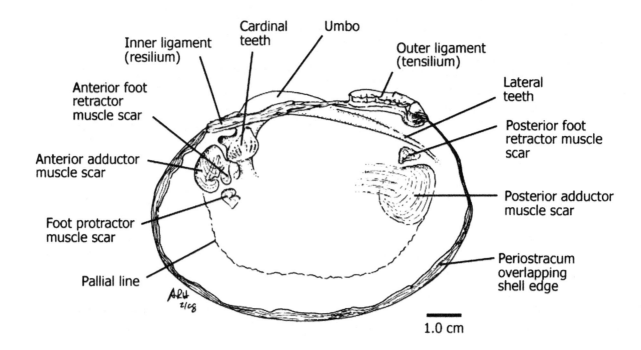

Figure 6.18. Internal anatomy of the right shell of *Unio*.

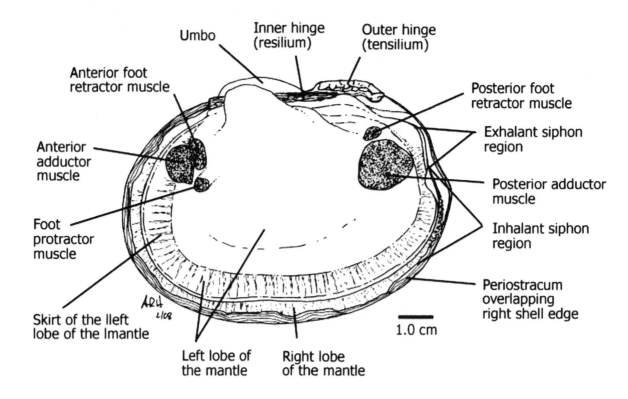

Figure 6.19. Soft tissues of *Unio* revealed by removing the left valve of the shell.

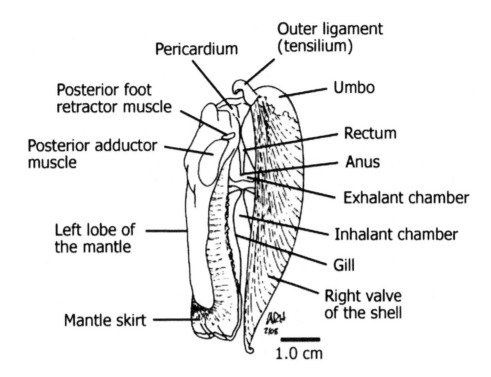

Figure 6.20. Posterior end of *Unio* with the left valve removed.

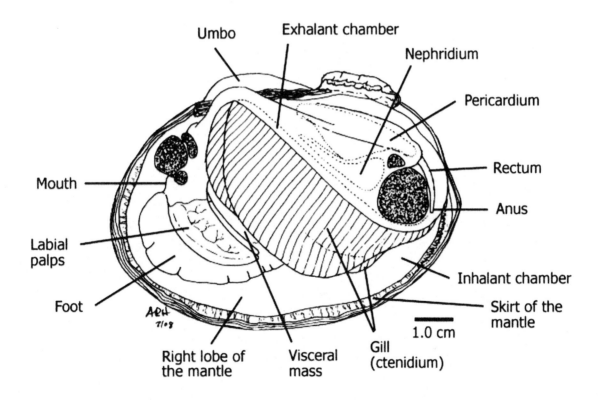

Figure 6.21. Anatomy of the mantle cavity of *Unio* with the left lobe of the mantle removed.

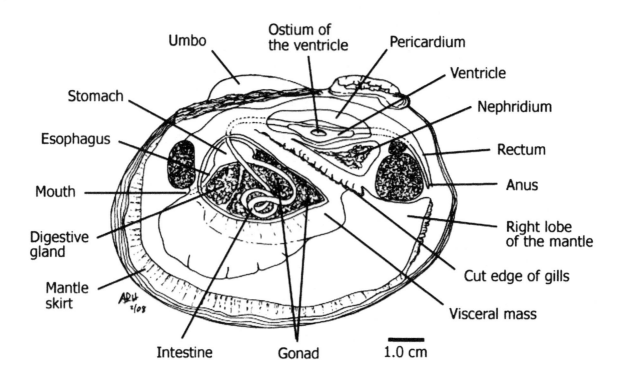

Figure 6.22. Internal anatomy of the visceral mass, pericardium, and nephridial sac of *Unio*.

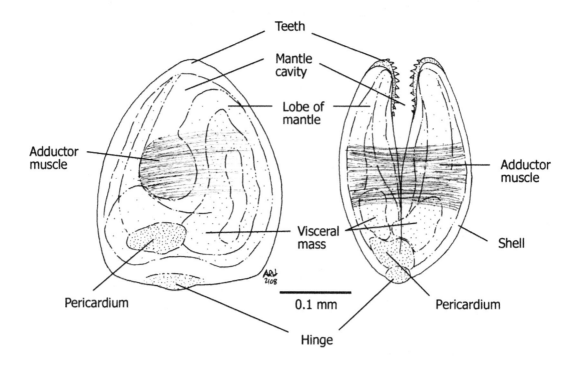

Figure 6.23. Freshwater mussel glochidia larva—lateral view (left).

Group Questions

1) How does the proportion of the body allocated to the digestive system in molluscs compare to what you saw in the body plans of the Platyhelminthes? Develop a hypothesis that explains the differences in allocation of body space to digestion in Molluscs and Platyhelminthes.
2) Develop a hypothesis that explains how the basic body plan of molluscs allowed them to become so diverse, and ecologically and evolutionarily successful.
3) Why do freshwater bivalves have glochidia larvae, while marine bivalves do not?

Chapter 7: Annelida

Clade Annelida includes about 20,000 living species of segmented worms. The name Annelida means "little rings", and refers to the ring-like appearance of segments of these animals' bodies.

Annelids have the following characteristics:

1) Metamerism (repeated structures in the body)
2) Segmented body
3) Prostomium (first body segment) and pygidium (last body segment)
4) Peristomium (second body segment) bears the mouth
5) Trochophore larva (in some)
6) The body wall from outer to inner layers includes a non-molted cuticle made of cross-helical fibers of the protein collagen, a mono-layered epidermis, a layer of connective tissue, a layer of circular muscle, a layer of longitudinal muscle, and a peritoneal layer that lines the pair of coelomic spaces in each segment of the body (except the prostomium that has only one coelomic space).
7) Chaetae made of beta-chitin (in most)
8) Closed circulatory system with many contractile vessels that function as "hearts"
9) Hemoglobin
10) Central nervous system containing regular sized and giant neurons.
11) Metanephridia
12) Nuchal organ (sensory)

Clade Annelida was traditionally divided into two groups, the Polychaeta and the Clitellata. Recent work on the taxonomy of Annelida rejects this taxonomy and replaces it with a new one. The new Clade Annelida includes Sipuncula, formerly an independent phylum, as a basal taxon within the clade, and sister taxon to a currently unnamed clade, "Clade 1". Clade 1 in turn includes two major taxa, Errantia and Sedentaria. Clade Errantia contains members of the old Class Polychaeta that have enlarged parapodia with prominent chaetae and anatomical characteristics supporting a motile, carnivorous lifestyle. Sedentaria, on the other hand, have reduced or no parapodia, and chaetae that are small and mainly embedded in the body wall. These animals make a living mainly by filter feeding or deposit feeding by ingesting sediment (think tubeworms and earthworms, oh and leeches too).

Sipuncula

The Sipuncula, commonly referred to as peanut worms, were an independent phylum until only a few years ago because they were considered to have a unique body plan, but as mentioned above, rigorous phylogenetic analysis now shows that sipunculids are a basal group of Clade Annelida. They are, however, the same animals they ever were, we just understand their evolutionary origins and connections better than we used to.

Sipunculids differ from other annelids in that they have a j-shaped gut, a retractable tentacle-bearing structure called an introvert, and they lack a segmented body and chaetae. They are similar to other annelids, however, in that they have trochophore-like pelagosphera larvae, a cross-linked cuticle made of collagenous fibers, a similar body wall, and nuchal organs.

Sipunculids are strictly marine, have a worm-shaped body, and are commonly found living in cracks, among cobbles, or as infauna in soft-sediment environments. The body includes an eversible structure called an introvert that bears the mouth, a j-shaped digestive tract with the anus located at the base of the introvert rather than terminally, and a two coelomic spaces, a large coelom in the trunk of the body and a smaller coelom in the tentacles of the introvert.

Tasks

1) Examine the external anatomy of your specimen. DRAW what you see.
2) Dissection. Immerse your specimen and make as incision that runs the entire length of the body. There are few obvious indicators of this animal's dorso-ventral polarity, but you will hopefully get lucky and be able to see paired structures such as the introvert retractor muscles and metanephridia. Be sure to look for the prominent longitudinal muscle bands that run the length of the body. Also note muscles associated with the intestine—the spindle muscle and radial fixing muscles. Look for the longitudinal nerve cord. It is lighter in color than the muscle bands of the body wall. Take particular care to study the anterior portion of the nerve cord and its relationship to the introvert and anterior region of the gut. You can usually see two large brownish sac-like structures just posterior to the introvert. These are the metanephridia. The gonads are easily detached from the body wall during initial stages of dissection, but if you are lucky you may see them still attached to the body wall near the base of one of the introvert retractor muscles. DRAW what you see. Refer to Figure 7.1 to help you identify what you see.

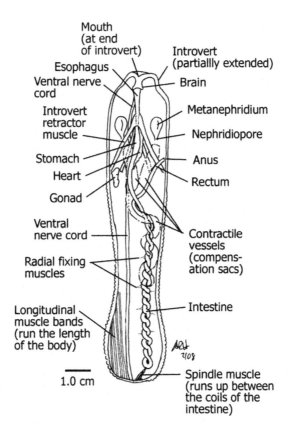

Figure 7.1. Internal anatomy of *Sipunculus nudis*.

Errantia

These annelids are mainly motile hunters. This exercise focuses on the common errant worm, *Nereis*. It is an active hunter that has strong grasping jaws and a large muscular pharynx.

Tasks

1) Examine the head of *Nereis*. This worm identifies its food and senses its environment mainly via chemical and tactile stimuli. Structures of the head region reflect that life style. Use a magnifying lens or dissecting scope examine your specimen. DRAW the head of your specimen. Refer to Figure 7.2 to help you identify what you see.
2) Examine parapodia. Use a magnifying glass or dissection scope to examine the parapodia associated with each body segment of *Nereis*. Use a compound scope to examine a prepared slide of a parapodium of *Nereis*. DRAW what you see. Refer to Figure 7.3 to help you identify what you see.
3) Study the internal anatomy of your specimen. Put your specimen of *Nereis* in a wax bottomed dissection tray, and add enough water to immerse the specimen. Starting at least 30 body segments back from the head, make a medial, longitudinal incision through dorsal body wall. The coelomic space of each segment is separated from neighboring segments by septae. Use a probe or forceps to cut through septae that connect the intestine and other organs to the body wall. Use insect pins to attach to opened body wall to the bottom of the dissection tray. Continue cutting and pinning the body wall as you move toward the head until you have exposed the anatomy of the entire anterior region of your specimen. DRAW what you see. Use Figure 7.4 to help you identify what you see.
4) Dissect the pharynx. Remove the pharynx and carefully open that organ. Look for a mass of muscles and the pincer-like jaws. These are hook-shaped and normally black in color. DRAW what you see.
5) Remove a section of the intestine so you can see the ventral blood vessel (dark tube/line) and the ventral nerve cord (light line). Add these structures to your drawing.
6) Look for a pair of metanephridia housed in the ventro-lateral areas of each body segment. Use a magnifying lens or dissection scope to do this. Nephridia can be delicate and are often destroyed when the body wall is opened. DRAW a metanephridium if you are able to locate one.
7) Use a compound scope to examine a prepared cross-section slide of *Nereis*. DRAW what you see. Use Figure 7.5 to help you identify what you see.

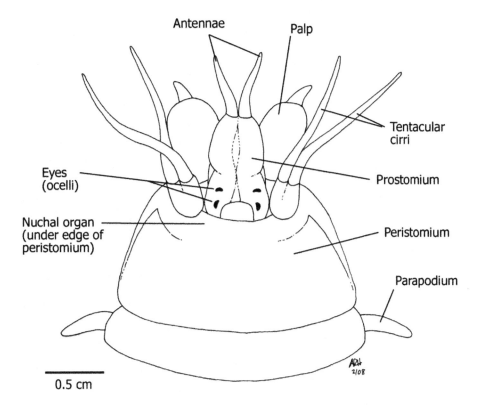

Figure 7.2. Structures of the head of *Nereis*.

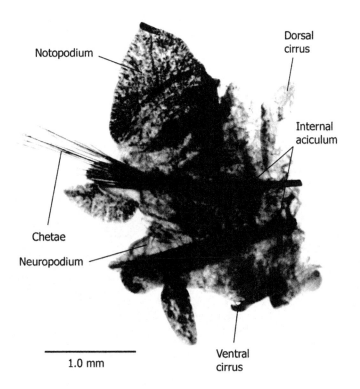

Figure 7.3. Anatomy of a parapodium from *Nereis*.

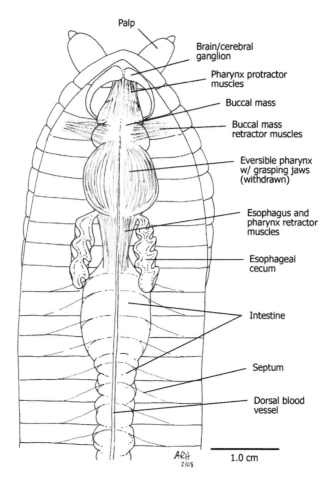

Figure 7.4. Internal anatomy of *Nereis*.

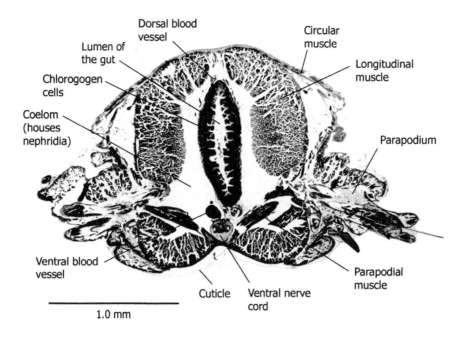

Figure 7.5. Cross-section through the intestinal region of *Nereis*.

Sedentaria

You will study two representatives of this clade: an earthworm, and a leech. Earthworms play an important role in nutrient cycling as they ingest soil and digest organic material from it. Their burrowing also aerates the soil, thus increasing the availability of oxygen to plant roots.

<u>Tasks</u>

1) Obtain a live earthworm and observe its behavior. Keep your specimen on a moist paper towel as you observe the contraction and elongation of individual body segments, and their collective effect during movement. Take notes on what you see.
2) Devise a simple experiment to see if earthworms prefer light or shade. Be sure to record the results of your experiment in your lab notebook.
3) Discover the location of chaetae on the earthworm *Lumbricus* by gently running your fingers along the length of the worm from anterior to posterior, and then from posterior to anterior. Repeat this for dorsal and ventral surfaces of the worm. Record your observations.
4) Place a live specimen on a dry paper towel and listen carefully as the worm moves. See if you can hear the scratching sound of chetae on the paper towel. Do not leave the worm on a dry paper towel, it will quickly desiccate.
5) Obtain a preserved specimen, place it in a wax-bottom dissection tray and add enough water to cover it. Use a magnifying lens or dissection scope and observe its external anatomy. DRAW what you see. Use Figure 7.6 to help you identify what you see. Be sure that structures on your drawing correspond to the correct number of segments from the prostomium.
6) Observe the internal anatomy of *Lumbricus* after opening the body wall by repeating the directions listed in Task #3 for *Nereis*. DRAW what you see. Refer to Figure 7.7 to help you identify what you see. You should be aware that you will not always see everything indicated on Figure 7.7 because the visibility of some structures depends on the reproductive state of your specimen.
7) Obtain a prepared cross-section slide of *Lumbricus* and observe it under a compound microscope. DRAW what you see, and use Figure 7.8 to help you identify what you see.
8) Obtain a preserved leech and use a magnifying glass or dissection scope to observe its external anatomy. Be aware that there appear to be many segments in their bodies, but what appear to be body segments are actually external annuli, usually over 90 of them, and each of these annuli do not correspond to true body segments. Leeches always have 33 internal body segments. DRAW what you see, and refer to Figure 7.9 to help you identify the external anatomy of *Hirudo*.
9) Body wall organization. Use a dissecting scope to examine the organization of fibers in the body wall of the leech. Look at both the dorsal and ventral body wall as you do this. DRAW the orientation of the fibers in the body wall.
10) You are not required to dissect the leech. This is a difficult dissection due to the reduced coelom and massive amount of muscle in the leech body. If, however, you have time and want to give this a try, follow the procedure to open up the body of the leech as you did for *Nereis* and *Lumbricus*. You may need to refer to figures in your textbook to inform you about what you see. Be advised that the body wall is thicker and more muscular than either worm you have looked at so far in this exercise. You can look for one pair of

ventro-laterally located metanephridia per body segment. Yellowish chlorogogen cells line the length of the intestine. DRAW what you see and identify everything you can.

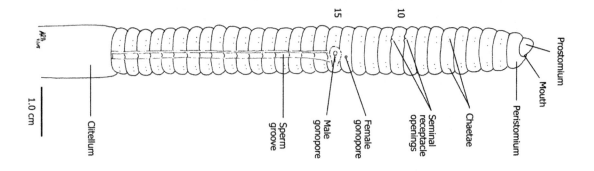

Figure 7.6. External anatomy of the earthworm *Lumbricus*. Segments 10 and 15 are indicated.

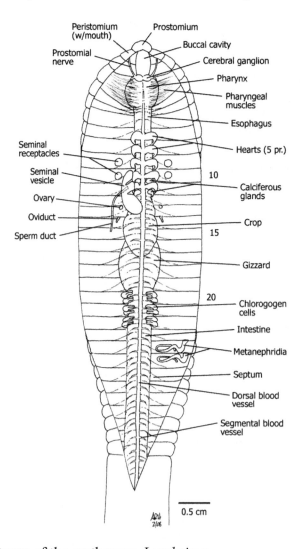

Figure 7.7. Internal anatomy of the earthworm *Lumbricus*.

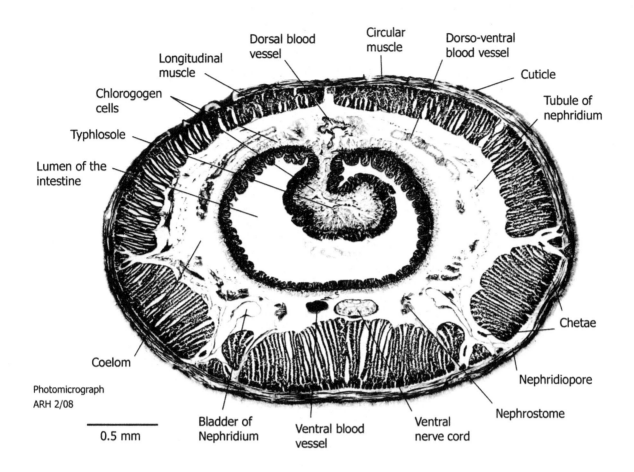

Figure 7.8. Cross section of the earthworm *Lumbricus* through the intestine.

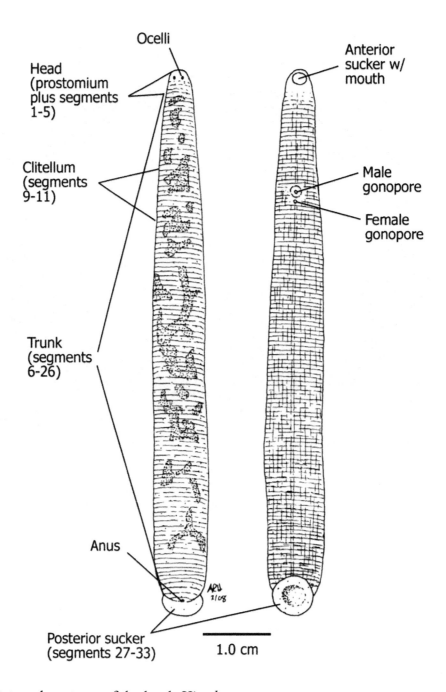

Figure 7.9. External anatomy of the leech *Hirudo*.

Group questions

1) What is the lifestyle of leeches? How is their anatomy adaptive to that lifestyle?
2) Most of you have seen earthworms in a biology lab before, but how has today's exercise expanded or changed your view of annelids?
3) At least some of the reproductive structures of *Lumbricus* are quite conspicuous. Where are the reproductive structures of Nereis?

Chapter 8: Brachiopoda and Nematoda

This may seem like an unusual pairing of taxa to cover in one lab, but time constraints of any academic calendar make a lab like this one inevitable. Even so, this pairing may not be that surprising considering what you have already done and what is coming up in lab. You just completed work on molluscs and annelids, both members of a larger clade called the Lophotrochozoa. Members of this clade use cilia for feeding and locomotion. There are many lophotrochozoan taxa that are not covered in this lab manual, but need to be introduced to at least one, the Brachiopoda.

The other taxon in this chapter is Nematoda, the roundworms. Roundworms and their close relations (none of which is covered in this chapter) – Nematomorpha, Loricifera, Kinorhycha, and Priapula – are members of Clade Cycloneuralia. Clade Cycloneuralia together with Clade Panarthropoda – Arthropoda, Tardigrada, and Lobopodia (Onychophora) – make up the larger Clade Ecdysozoa. Ecdysozoans lack locomotory cilia, but they have a tough chitinous outer cuticle that is molted by members of some groups via a specific process called ecdysis. Panarthropoda is the topic of the next chapter.

Brachiopoda

Brachiopods are extremely important index fossils. Index fossils are species that had a large geographic range, produced many fossils, and are present in the fossil record for only a limited time before they were replaced by other species. Brachiopods were once the dominant shelled, suspension-feeding taxon. Bivalve molluscs have since replaced them, but Brachiopods still survive in refuge habitats.

Brachiopods and all lophophorates bear a crown of ciliated tentacles called a lophophore. This is the organ they use for feeding and is the primary site of gas exchange. Lophophore size and shape varies from species to species. This part of the chapter introduces you to two species of brachiopods. The first one, the inarticulate brachiopod *Lingula*, lives an infaunal life in soft sediments. This is the oldest known living species of animal. Fossils of *Lingula* from the Cambrian Era, over 500 million years ago, are virtually indistinguishable from living specimens. The other one is the articulate brachiopod *Terebratalia*. By the way, all brachiopods have two shells, usually made of potassium phosphate, and are dorsal and ventral shells, not left and right-hand shells as you saw in bivalves. Inarticulate brachiopods do not have a hinge connecting their shells, while articulate brachiopods do.

<u>Tasks</u>

1) Use a magnifying lens or dissection scope to study the shell and external anatomy of *Lingula*. Detach the pedicle from the shell and make a cross-sectional cut through it and examine its anatomy. DRAW this specimen's external anatomy. Refer to Figure 8.1 to identify what you see.
2) Unlike bivalves which have a pair of lateral shells, brachiopod shells are dorso-ventral in orientation, and you can't tell which is dorsal until you separate the shells. Use a probe tip to separate the mantle and anterior adductor muscles from the inside of the shell. Gently pull the shells apart. Doing so will expose structures of the mantle cavity and viscera. Or, you may choose to use forceps to crack and peel away small pieces of the

shell until the structures inside are exposed. Note: The lophophore is always attached to the dorsal valve of the shell. Once you know which shell is dorsal and which is ventral, remove the ventral shell, exposing the lophophore, muscles, and other soft tissues. Doing this also exposes the mantle cavity and the mantle. Branches of the coelom extend to near the margin of the mantle. DRAW what you see. Refer to Figure 8.2 to help you identify what you see.

3) The posterior portion of the shell houses the viscera. By now you have enough experience that you should be able to identify the greenish digestive cecum, yellowish gonads, and orange-red lateral blood vessels. Locate the anterior and posterior adductor muscles. The anterior adductor muscles are divided into two parts, catch muscle fibers that contract quickly, and fibers that contact slowly. These muscles can remain contracted for prolonged periods of time and are used to keep other animals from prying the shells apart. Also note the three sheet-like oblique adjustor muscles. These hold the shell valves in correct position relative to each other. DRAW what you see. Look at Figure 8.2 to help you identify what you see.

4) Remove the exposed digestive cecum, gonads, lateral vessels, anterior adductor muscles, and oblique adjustor muscles. Doing so will uncover structures of the digestive tract and the anterior adjustor muscles DRAW the digestive tract and other exposed soft tissues, and look at Figure 8.2 to help you to identify what you see.

5) Now move on to the articulate brachiopod *Terebratalia*. Examine the external anatomy of your specimen. DRAW the shell of *Terebratalia*. Look at the top drawing in Figure 8.3 to help you identify what you see.

6) Insert a scalpel blade or probe tip between the dorsal and ventral shell valves of your specimen and pry them apart far enough to use your fingers to gently pull them apart. Pull on the dorsal shell valve until you feel the hinge start to give way, and then carefully separate the shell valves from each other. This exposes the structures in the mantle cavity. Immerse your specimen. Take time to observe the structures associated with the ventral valve as well as the lophophore (which is attached to the dorsal valve) and other structures associated with the dorsal valve. DRAW what you see, and look at drawings in Figure 8.3 to help you identify what you see.

7) Digestive tract. Remove the lophophore and adductor muscles from *Terebratalia*. This exposes structures of the digestive tract. DRAW what you see and use Figure 8.3 to help you identify what you see.

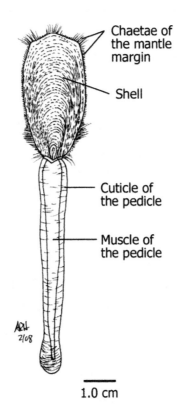

Figure 8.1. External anatomy of the inarticulate brachiopod *Lingula*.

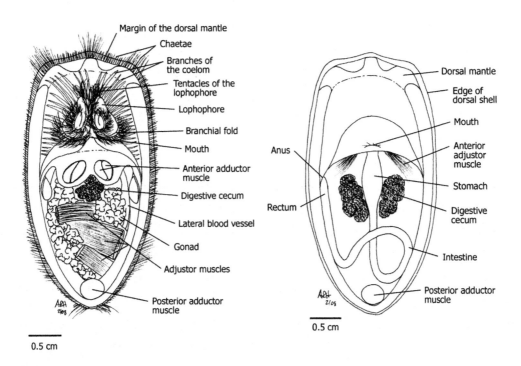

Figure 8.2. Internal anatomy of *Lingula*: lophophore and superficial soft tissues, left; digestive tract and deep tissues, right.

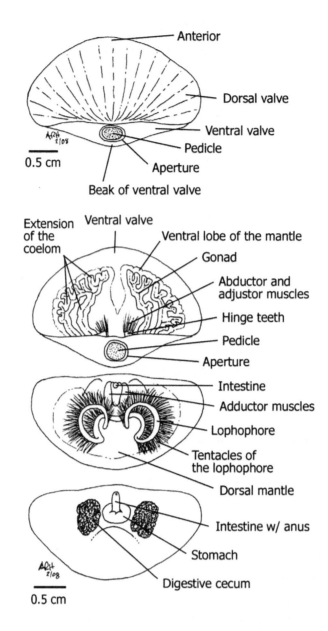

Figure 8.3. External and internal anatomy of the articulate brachiopod *Terebratalia*: Shell, top; mantle and branches of the coelom, upper middle; lophophore and superficial soft tissues, lower middle; digestive tract and deep soft tissues, bottom.

Nematoda

Nematodes, the roundworms, are almost certainly the most numerically abundant taxon of animals on the planet. We know a lot about a few kinds of nematodes because they are medically important human parasites, but there are vast numbers of species and incalculable numbers of individuals about which we know nothing. This makes estimating their diversity extremely difficult. Most nematodes live in the soil or other habitats where they appear to be benign with regard to human health.

Nematodes, like arthropods, molt their cuticle as they grow, except that nematodes do this only a few times very early in their life, while arthropods carry out ecdysis periodically throughout their lives. Nematodes have a relatively simple anatomy. They have a tough outer cuticle that is made of beta-chitin, four blocks of longitudinal muscles, and, depending on the size of the worm, either a rather spacious pseudocoelom to none at all. A pseudocoelom is a fluid-filled space in the body that is lined on its outer edges by mesodermal peritoneum, but its inner surface is lined by endodermal tissue of the gut. Many nematodes also exhibit a trait called eutely. This means that in some cases they produce only a predetermined number of cells for an organ or for an entire body. This is the trait that makes the worm *Caenorhabditis elegans* such a useful species for doing developmental biology.

In this exercise you have a chance to examine one nematode in detail, and to take a peek at some other representatives.

Tasks

1) Behavior. Acquire some live specimens of the vinegar eel *Tubatrix aceti*. Place them in a small glass bowl with enough vinegar so they can move freely. Use a dissection scope to observe their behavior. Be sure to include your observations in your lab notebook.
2) External anatomy of the hog intestinal roundworm *Ascaris lumbricoides*. Work with a lab partner during this section of the lab—and be sure to glove up, because female *Ascaris lumbricoides* always contain thousands of viable embryonated eggs, even if they have been immersed for prolonged periods of time in toxic fluids like formalin. *Ascaris lumbricoides* is one of the largest roundworms, reaching lengths of over 30 cm. Obtain preserved specimens of female and a male. Females tend to be longer and larger than males, and females are pointed at both ends. Males, on the other hand, are smaller and the body forms a pronounced hook at the posterior end. Use a magnifying lens or dissection scope to locate the three lips at the anterior end of the body. *Ascaris* have one dorsal lip and two ventro-lateral lips. You can determine which the dorsal lip is by finding the lateral epidermal cords that run along the sides of the body of *Ascaris*. The lip that is medial relative to the lateral cords is the dorsal lip. Once you have identified the dorsal surface, don't lose track of it. Use insect pins to attach your worm, ventral surface down, to a wax bottomed dissection tray. DRAW the external anatomy of your worms.
3) Internal anatomy of *Ascaris*. One of you should dissect a female and the other a male worm. Show each other what you see as you work through these dissections. Use a scalpel to open the body cavity by making a longitudinal, medial incision through the dorsal body wall. Be careful as you make the incision, because the body wall is thin and the structures inside the body cavity are delicate. Fold the body wall back and use insect pins to attach the body wall to the bottom of your dissection pan. If you are dissecting a female continue your incision for the entire length of the body. If you are dissecting a male, stop your incision short of the hooked posterior. DRAW the internal anatomy of your specimen after you have the body wall of your specimen pinned down. You may need to gently tease apart some structures in order to see them and their spatial relationships to each other. Look at Figures 8.4 and 8.5 to identify what you see.
4) Eggs. Make a wet-mount slide of contents of the uterus of the female work. Infertile eggs have an outer covering that looks like it contains many small droplets. A fertile egg

contains a single large cell with a visible nucleus. Some egg capsules may contain embryonic worms. DRAW what you see.

5) Examine the male reproductive system. Cut the hooked end off of the male worm. Make a longitudinal cut directly through that structure as shown in Figure 8.5. Use a dissection scope to see if you can identify structures of the male cloacal region.

6) Use a compound scope to examine a prepared cross-section slide of *Ascaris*. Study cross-section slides of male and female *Ascaris* made through the intestine, as well as a cross-section slide through the pharynx of *Ascaris*. DRAW what you see, and look at Figures 8.6-8.8 to help you identify what you see.

7) If time allows, look at prepared slides of medically important nematodes. Be sure to DRAW what you see.

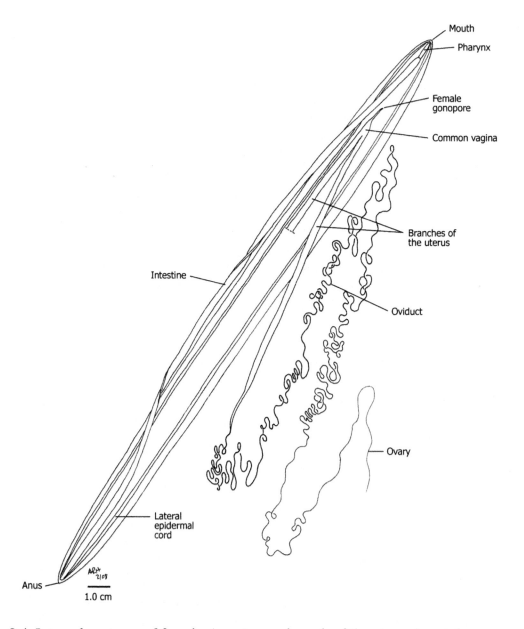

Figure 8.4. Internal anatomy of female *Ascaris*; one branch of the uterus is not drawn.

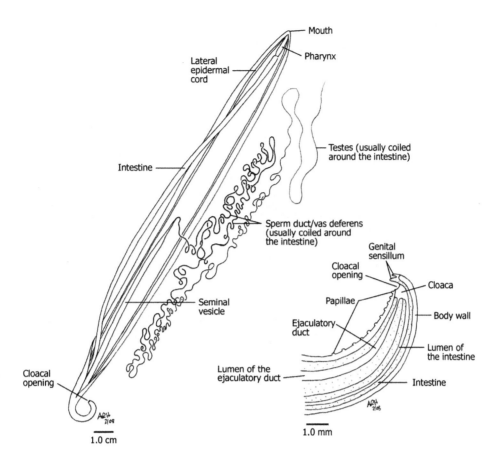

Figure 8.5. Internal anatomy of male *Ascaris* and detail of the clocal region of male *Ascaris*, bottom right.

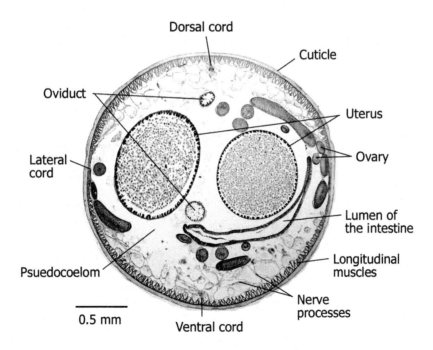

Figure 8.6. Cross-section slide through the intestine of female *Ascaris*.

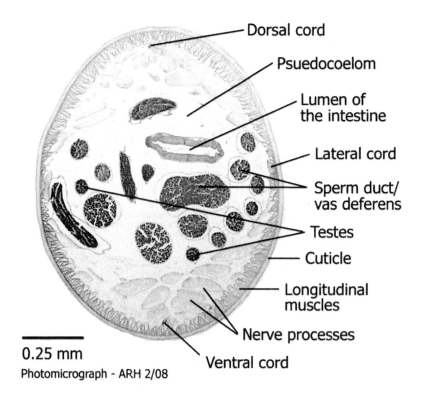

Figure 8.7. Cross-section slide through the intestine of male *Ascaris*.

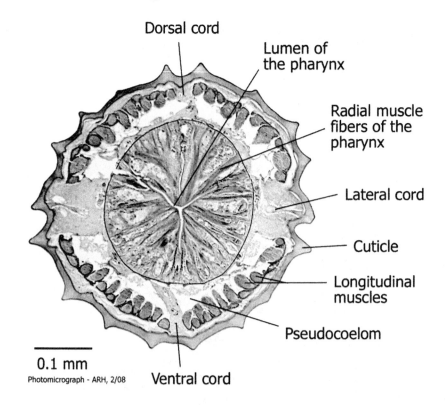

Figure 8.8. Cross-section slide through the pharynx of *Ascaris*.

Group Questions

1) How can nematodes can get along with only longitudinal muscles in their body wall?
2) Compare and contrast the body plans of parasitic roundworms and parasitic flatworms.
3) Brachiopods were once extremely abundant, both numerically and taxonomically. Today, however, there are only about 350 living species of brachiopods, compared to 12,000 known fossil species. It has been suggested that the evolution of bivalve molluscs led to the decline in the dominance of brachiopods in the fossil record. Develop a hypothesis that explains how the evolution of bivalves caused a decline in brachiopod biodiversity and geographic distribution.

Chapter 9: Panarthropoda

Clade Panarthropoda includes Lobopodia, velvet or walking worms; Tardigrada, water bears; and Arthropoda.

Members of Panarthropoda have the following characteristics:

1) Segmented body
2) Paired appendages
3) A non-elastic outer cuticle that must be molted to allow increase in body size

There is enough material in this chapter for at least two lab meetings.

Lobopodia

Living members of the Lobopodia are called Onychophorans, commonly known as velvet worms or walking worms. There are just over 100 living species of Onychophorans. They live in moist tropical or subtropical regions because high humidity is vital to these animals. Why? They cannot dilate the spiracles of their tracheal systems.

Task

1) Use a magnifying glass or dissection scope to examine a preserved onychophoran. Be sure to examine the anatomy of the head, dorsal and ventral, lobopods (appendages) of the body, and structures of the posterior end of the body. DRAW what you see. Refer to Figure 9.1 to help you identify what you see.

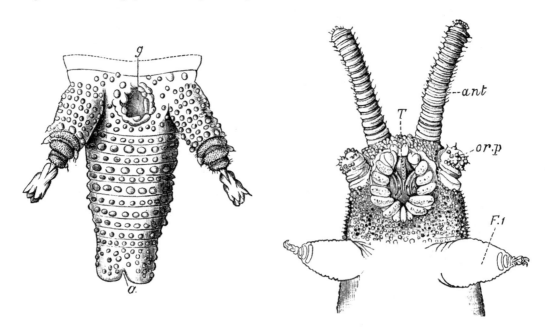

Figure 9.1. External anatomy of the posterior, left, and anterior, right, ends of an onychophoran: a, anus; ant, antenna; F1, first leg; g, gonopore; or p, oral papilla; T, tongue (Public domain image, by Sedgewick in The Cambridge Natural History, 1922, V p. 6).

Tardigrada

There are about 1000 described species of Tardigrades, commonly called water bears or moss piglets, but they constantly go unseen and unappreciated because they are so small. Once you see them, however, they are likely to become a personal favorite. Tardigrades are microscopic, and live in marine, freshwater, and semi-terrestrial environments. The largest known tardigrade is 1.5 mm long, and most species are only 100-150 micrometers long.

If you want to see these you have to go on tardigrade safari and collect materials where they are likely to be found. Promising materials include mosses growing in seldom-cleaned rain gutters, bird baths, finely branched aquatic plants, and sandy marine substrates.

<u>Tasks</u>

1) Collect materials where tardigrades may exist. These include things like moss from moist or even dry rain gutters, finely branched aquatic plants or algae, or marine sediments. Soak moss or aquatic plants in 5-10% ethyl alcohol solution. After 10-20 minutes in the solution, gently shake the moss or aquatic plant. If there are tardigrades present they will be completely "relaxed" by now and should release their hold and drift to the bottom of the container. Use a Pasteur pipette to collect samples from the bottom of the jar or bowl, and make wet mounts of your samples. Use a compound scope to look for tardigrades. You are likely to see all kinds of things in your sample, but keep an eye out for tardigrades, they are identifiable because they are the only thing in there with lobe-shaped appendages that bear chitinous hooks, as shown in Figure 9.2. If you find one, or think you found one, get excited, check with your instructor, show your neighbor, and then DRAW what you see.

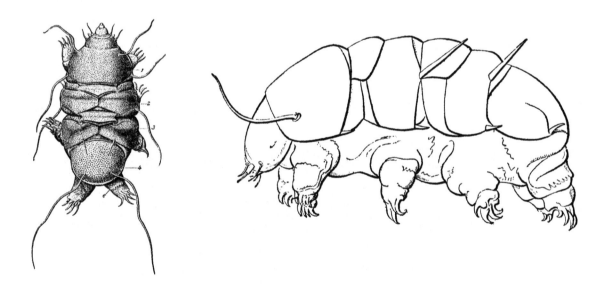

Figure 9.2. Tardigrades, *Echiniscus testudo*, left, *E. spinulosus*, right (Public domain images, from Doyere in The Cambridge Natural History, 1920, IV, pp. 478, 479)

Arthropoda

There are well over one million described species of Arthropods. Clade Arthropoda is in terms of number of species and almost certainly by ecological significance the most successful group of animals on Earth. They occupy virtually every life-supporting habitat on the planet including marine, freshwater, and terrestrial environments. Arthropods are assigned to the following taxa: Trilobitomorpha (extinct), Chelicerata, Crustacea, Myriapoda, Hexapoda, and a number of smaller groups not addressed here.

Arthropods share the following characteristics:

1) Chitinous exoskeleton
2) Jointed appendages
3) Segmented body
4) Tagma, the fusion of multiple segments to create specialized body parts such as a head, thorax, abdomen, post-abdomen, etc.

Trilobitomorpha

While we spend almost all of our time examining living taxa in this lab manual, you need to take at least a little time to look at trilobites because they are such an important group in the fossil record. They, like the brachiopods, were so common, diverse, and geographically widespread in the marine environment that we also use them as index fossils.

Task

1) Examine one or two trilobite fossils and DRAW what you see. Look at Figure 9.3 to help you identify what you see. I also encourage you to visit a geology or natural history museum if possible where you can see more trilobites as well as other invertebrate fossils. This visit probably needs to be made outside of class.

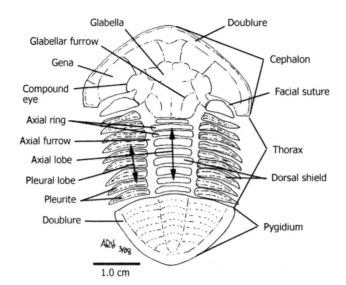

Figure 9.3. Trilobite, dorsal view.

Chelicerata

There are about 78,000 described species of Chelicerates. Over 90% of chelicerates are terrestrial, including scorpions, spiders, and their relations, and there are about 1300 species of pycnogonids – sea spiders.

Chelicerates share the following traits:

1) A body with two tagma: a cephalothorax, sometimes called a prosoma, and an abdomen. Ancestral forms also have a post-abdomen.
2) The cephalothorax bears six pairs of appendages.
    a. The first pair of appendages are the chelicerae, usually a grasping appendage.
    b. The second pair of appendages are the pedipalps. These can be used for sensory or even signaling functions.
    c. There are four pairs of walking legs.

This part of this exercise covers two taxa of the Chelicerata: Xiphosura, horseshoe crabs; and Arachnida, spiders, scorpions, ticks, etc.

Xiphosura

Your first stop is a look at a horseshoe crab. Horseshoe crabs, like the brachiopod *Lingula*, are living fossils. Horseshoe crabs appeared in the fossil record during the early Silurian period, over 400 million years ago, that's long before dinosaurs showed up. There are only a few surviving genera of Xiphosura. You will study the external anatomy of *Limulus*, a horseshoe crab commonly found along the east coast of North America.

<u>Tasks</u>

1) Study the external anatomy of the dorsal and ventral surfaces of *Limulus*. DRAW what you see. Refer to Figure 9.4 to help you identify what you see.
2) Use a dissection or compound microscope to examine a prepared slide of the trilobite larva of *Limulus*. DRAW what you see, and look at Figure 9.5 to help you identify what you see.

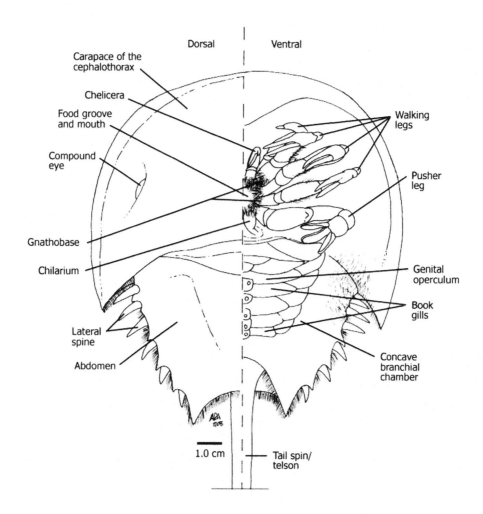

Figure 9.4. External anatomy of *Limulus,* dorsal, left; and ventral, right.

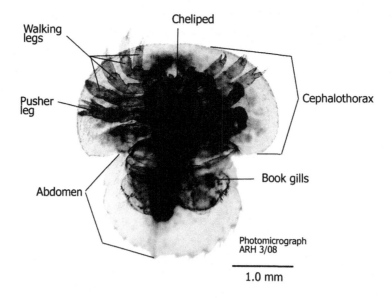

Figure 9.5. Trilobite larva of *Limulus*.

Arachnida

Clade Arachnida is a large and diverse taxon that lives mainly in terrestrial environments. Living members of this group include spiders, scorpions, ticks, and their relatives. By the way, of the three body plans, scorpions exhibit the most ancestral traits, and ticks the most derived traits, with spiders someplace in the middle.

Tasks

Team up with a partner. Between the two of you, you need to cover the following three specimens. After you have studied your selected specimen, show and teach your partner what you learned.

1) Scorpion. Study the external anatomy of a preserved scorpion. Use a dissection scope or magnifying lens as needed. DRAW what you see. Refer to Figure 9.6 for assistance in identifying structures you see.
2) Spider. Study the external anatomy of a preserved spider. Use a dissection scope or magnifying lens as needed. DRAW what you see. Refer to figures in your textbook for assistance in identifying structures you see.
3) Tick. Study a prepared slide of a tick. Use a dissection scope, dissection scope, or magnifying lens as needed. DRAW what you see. Refer to figures in your textbook for assistance in identifying structures you see.

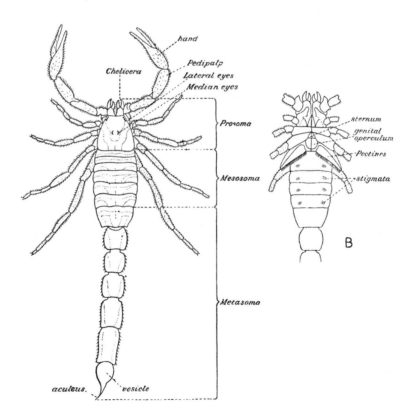

Figure 9.6. External anatomy of the scorpion *Buthus occitanus*: dorsal view, left; ventral view, right. (Public domain image, after Kraepelin, in The Cambridge Natural History, 1920, IV, p. 302.)

Crustacea

The Crustacea is another large group of arthropods with more than 50,000 described living species. This is a particularly important group because it includes many species that we use for food, e.g., lobsters, crabs, shrimp, and crayfish. It also includes barnacles, krill, and others. Though largely aquatic, there are a few terrestrial crustaceans. You probably already know at least one. It has many common names including woodlouse, pill bug, potato bug, armadillo bug, sow bug, rolly-polly bug, and a personal favorite of mine, chucky pig (used in Devon, England). Crustaceans, wherever they live, play important ecological roles as predators, herbivores, omnivores, filter-feeders, scavengers, and even parasites.

Crustaceans have the following traits:

1) Typically two tagma: cephalothorax and abdomen
2) Many specialized appendages including
    a. First appendage is a pair of antennules
    b. Second appendage is a pair of antennae
    c. Third appendage is a pair of crushing or cutting mandibles
    d. Two pairs of maxillae
    e. Maxillipeds
    f. Pereopods (walking legs)
    g. Pleopods (swimmerettes)
3) Nauplius larva (in many)
4) Tripartite brain
5) Saccate nephridia
6) A cuticle with four layers

Tasks

During this lab you will study the body plan of a crayfish, the blue crab *Callinectes*, and the water flea *Daphnia*.

Daphnia

1) Examine a prepared slide of *Daphnia* using a compound scope. DRAW what you see. Use Figure 9.7 to help you identify what you see.

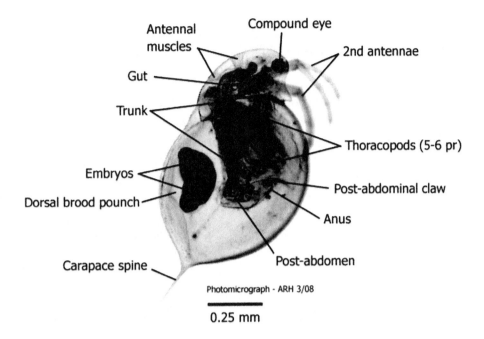

Figure 9.7. Anatomy of the water flea, *Daphnia*.

Pair up for the rest of this lab. One partner will lead the investigation of the crayfish and the other one will lead the investigation of the blue crab.

Crayfish

1) Obtain a preserved crayfish and immerse it. Use Figure 9.8 to determine its gender.
2) Examine the external anatomy of the crayfish. DRAW what you see.
3) Expose the gills of the crayfish. Cut through the exoskeleton starting at the posterior edge of the carapace just off-center of the dorsal mid-line. Continue that cut until you reach the rostrum. Next, make a cut from the leading edge of the rostrum ventrally until you reach the lower edge of the carapace. Carefully insert a probe between the carapace and soft tissues of the body, and separate any muscle attachments or other connections between soft tissues and the carapace. Remove the carapace. DRAW your specimen, focusing on the gill chamber. Use Figure 9.9 to help you identify what you see.
4) Examine the specialization of crayfish appendages. Review Figure 9.10 before you remove any appendages. Use a pair of forceps to carefully remove all of the appendages from one side of your specimen. Be sure that you remove the entire appendage, including gills that are exopods of several of the appendages. Lay the appendages out in order from anterior to posterior on a paper towel or in a dissection tray. Refer to Figure 9.9 again to help you identify each appendage. LEARN the different kinds of appendages, their names, locations on the body, and main functions.
5) Examine the internal anatomy of the crayfish. Remove the thin inner wall of the gill chamber and note the large lateral digestive cecum. You may see the long circumesophageal connective nerves passing around the esophagus in the anterior part of the body cavity. Remove the digestive cecum and look for structures of the digestive

tract and reproductive system. DRAW what you see. Refer to Figure 9.9 to help you identify what you see.

6) After you have identified and sketched structures of the cephalothorax, cut away the exoskeleton from the dorsal half of the abdomen. You should now be able to see the intestine running along a dorsal groove in the abdominal muscles. Next, carefully remove the stomach and cut it in half by making a dorsal incision along the midline. Look for the teeth of the gastric mill, as well as for the cardiac and pyloric chambers. See your textbook for additional detail on the stomach of crayfish.

7) Remove the muscles of the abdomen. Look for the whitish ventral nerve cord and segmental ganglia that lie along the ventral wall of the exoskeleton in that part of the body. Look also in the rostrum to locate the brain. Again, keep your lab partner filled in on what you see.

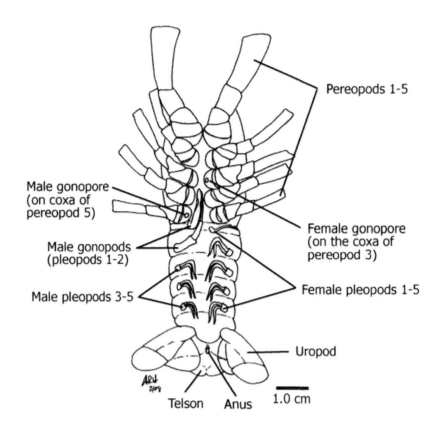

Figure 9.8. Ventral view of a crayfish, showing differences between female (left side) and male (right side) swimmerettes.

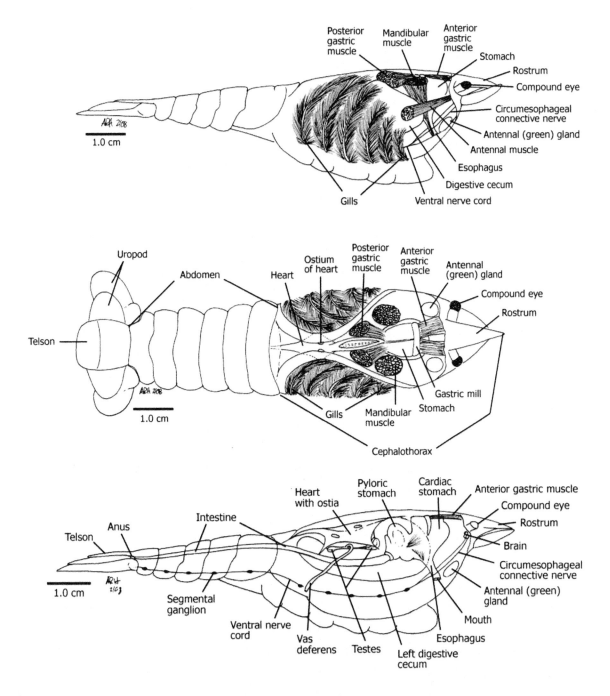

Figure 9.9. Internal anatomy of the crayfish. Top and middle drawings show anatomy of the gill chamber and superficial soft tissues of the body cavity. The lower image shows soft tissues exposed when gills and digestive caecum are removed.

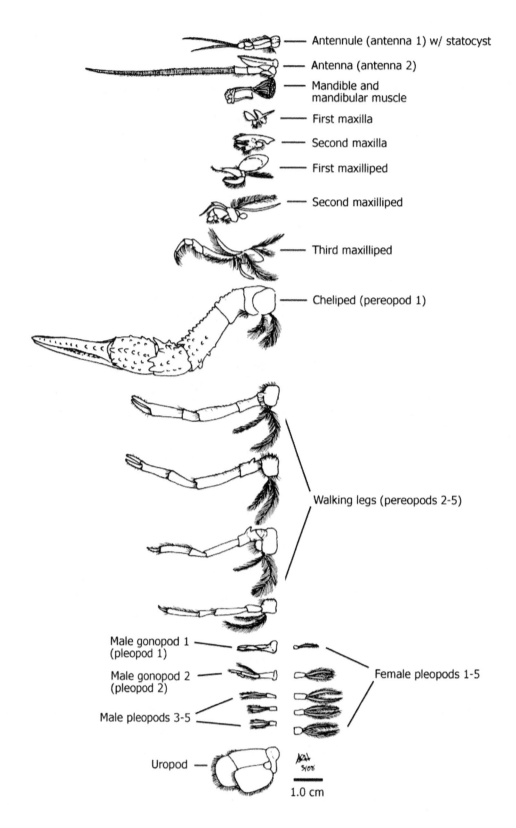

Figure 9.10 Appendages of the crayfish, from anterior at the top to posterior at the bottom. Note the sexual dimorphism of pleopods.

Callinectes, the blue crab

1) Obtain a blue crab and immerse it. Determine the gender of your specimen by looking at the shape of the abdomen. If the abdomen is broad then you have a female, as shown in Figure 9.11. If the abdomen is narrow, the specimen is male.

2) External anatomy. DRAW the external anatomy of your specimen. Refer to Figure 9.11 to help you identify what you see. Note: Crab pleopods are not shown on Figure 9.11. They can be exposed by inserting a probe between the tip of the abdomen and the plates of the sternum on the ventral surface of the body, and gently pulling the abdomen away from the cephalothorax. Pleopods of females are long and feather-like, while pleopods of males are long, cylindrical, and lack setae. Male pleopods are specialized for transferring spermatophores to the female, and female pleopods for holding eggs.

3) Gills. The gills are external to the body cavity but they are housed in a protective gill chamber formed by structures of the exoskeleton. The gills are exopods of appendages. The gills are exposed when you open the body cavity in the next step of dissection.

4) Open the body cavity by cutting away the dorsal wall of the carapace. The best way to do this is to insert the tip of a pair of scissors into the suture line between the posterior edge of the carapace and the first segment of the abdomen. Carefully cut from that point all the way around the dorsal margin of the carapace, cutting as close to the outer edge of the carapace as possible. Once you have completed this cut use a probe to separate any soft tissues or muscle attachments from the exoskeleton as you gently lift and remove the dorsal portion of the carapace. If you are not careful you could rip the heart and dorsal wall of the stomach as you remove the carapace. DRAW what you see. Look at Figure 9.12 to help you identify what you see. If the circulatory system of your specimen is injected with latex you may need to remove some excess latex before proceeding.

5) Once you have identified and drawn the exposed structures, remove the digestive cecum and ovary or testes from one side of the body cavity. The testes are in the same location as the ovaries on Figure 9.12. Doing this makes it easier for you to find the gill cleaner. Once you have done that remove the heart and the portion of the digestive cecum that extends ventral to below the heart. You should now be able to see the intestine and, in the female, a pair of large oblong seminal receptacles. Lastly, remove the stomach and then cut it in half by making an incision through the dorsal midline. Look inside the stomach for teeth of the gastric mill, as well as for the cardiac and pyloric chambers of the stomach. Refer to your textbook for additional information on the anatomy of the stomach. Stomachs of crabs and crayfish are similar.

6) Appendages. Before you remove any appendages review Figure 9.10 that shows the order and types of appendages found on crayfish. Crabs have the same type and number of appendages. Use a pair of forceps or your hands to carefully remove all of the appendages from one side of your specimen. Be sure that you remove the entire appendage, including gills if possible. This is easier in crayfish than crabs, but do your best. Lay the appendages out on a paper towel, in order from anterior to posterior. Refer to Figure 9.10 again to help you identify each appendage. LEARN the different kinds of appendages, their names, locations on the body, and main functions.

7) Nervous system. Carefully remove all remaining organs from the body cavity of the crab. Look up into rostrum region of the body cavity to locate the brain, ventral nerve cord, and other structures of the nervous system.

8) Developmental stages of crustaceans. Study prepared slides of the barnacle nauplius, crab zoea, and crab megalops larval stages. Refer to Figure 9.13 to help you identify what you see.

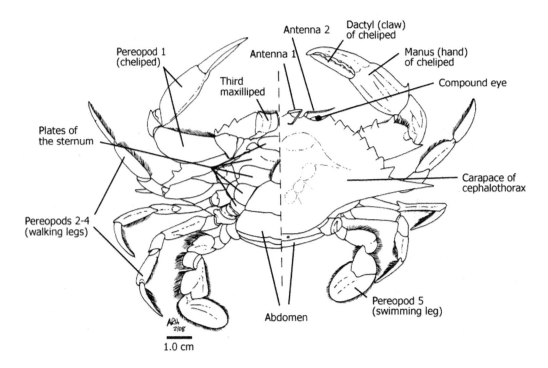

Figure 9.11. External anatomy of *Callinectes*: ventral view, left; dorsal view, right.

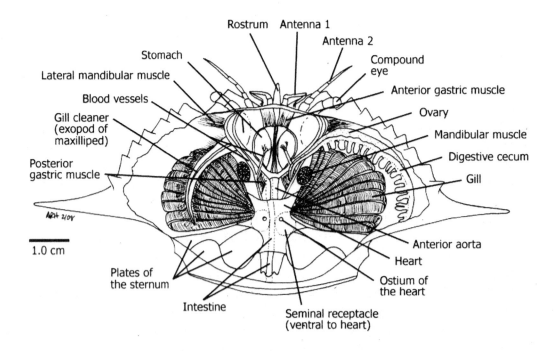

Figure 9.12. Gill chambers and internal anatomy of *Callinectes*. The digestive cecum, ovary, and associated structures are not shown on the left side of the drawing.

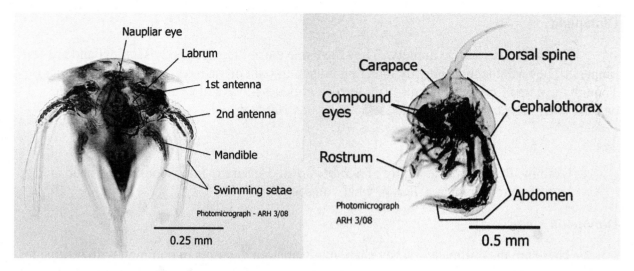
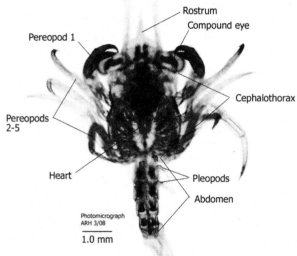

Figure 9.12. Crustacean larval stages: nauplius. upper left; zoea, upper right; megalops, below.

Myriapoda

Myriapods include the Chilopoda, centipedes, and Diplopoda, millipedes, and some smaller groups not included in this chapter. There are, all together, about 13,000 species in this group, all terrestrial.

All Myriapods have the following traits:

1) Many identical body segments
2) Each segment has a dorsal tergite and ventral sternite (thick plates of the exoskeleton)
3) Trachea for gas exchange
4) Malpighian tubules for excretion
5) A single pair of antennae
6) Tomosvary organs for sensing humidity

Chilopoda

Centipedes are active hunters. They have one pair of legs per body segment and are fast runners. They subdue their prey by injecting poison into them via a pair of forcipules (fangs). Though the name Centipede means "100 feet" centipedes always have an odd number of leg-bearing body segments, so no centipede actually has 100 feet.

Tasks

1) Examine the external anatomy of *Scolopendra*. Be sure to take a look at the head and legs of your specimen. DRAW what you see.

Diplopoda

These are the millipedes. They share many anatomical traits in common with centipedes, but millipedes are not carnivorous. Instead they make a living by eating living or dead plant matter. The first three body segments have only one pair of legs per segment, but the rest have two pairs of legs per body segment.

Tasks

1) Examine the external anatomy of *Spirobolus* or other available millipede. Be careful with these specimens, since they tend to be quite fragile. Take a close look at the head and at the legs on body segments of your specimen. DRAW what you see.

Hexapoda

This is the largest clade in Arthropoda. It contains over a million described living species; these are the insects. There are more kinds of insects than all other species of living things combined, and the vast majority of these are beetles. Insects play important roles in the ecology of our planet. Huge numbers of plants are pollinated only by insects, and other insects are significant crop pests. They also play important roles as disease vectors, decomposers, scavengers, predators, etc.

You may be somewhat disappointed at the superficial treatment of the Hexapoda in this chapter, but remember that the goal of this lab manual is to introduce the diversity of body plans, not the biodiversity within each body plan.

Hexapods share these traits:

1) Three body tagma: head, thorax, abdomen
2) Two pairs of wings, always on the thorax
3) Three pairs of legs, always on the thorax
4) One pair of antennae

Grasshopper

1) Examine the external anatomy of your specimen. Use a magnifying lens or dissection scope to look for spiracles and mouthparts. DRAW what you see.

2) Use a pair of forceps to remove one set of appendages. Leave the coxa (basal article) of walking legs attached to the body. If you try to remove the coxa with the rest of the leg it will pull muscles from inside the body cavity, and that could mess up what you hope to see when you study grasshopper internal anatomy. LEARN the names and functions of each appendage.

3) Examine the internal anatomy of the grasshopper. Do not rush as you carry out this dissection. First remove the wings and any walking legs that are still attached to the body (leave the coxa of all legs in place). Use scissors to cut around the edges of the pronatum, and use a probe to scrape the inner surface of the pronatum as you remove it. This separates any muscle attachments to the pronatum. Next, use scissors to make a longitudinal cut through the dorsal wall of the exoskeleton that runs the entire length of the thorax and abdomen. Make short lateral cuts along the length of the body and pin the exoskeleton to the floor of a wax bottomed dissection pan. Add enough water to immerse your specimen. You should see a thin layer of reddish tissue covering much of the internal body cavity. That reddish tissue includes longitudinal and circular abdominal muscles. Running along the dorsal surface of that tissue you may also see the dorsal vessel/heart. After you have made a few observational notes use forceps (not scissors!) to carefully remove the reddish tissue layer. DRAW what you see. Refer to Figure 9.13 to help you identify what you see.

4) Structures of the reproductive system are in the abdomen, dorsal to the digestive tract. The reproductive organs are paired structures that lie mainly along the dorsal surface of the mid and hindguts of the digestive tract. Remove one set of the reproductive organs to expose structures ventral to them. DRAW what you see.

5) The nervous system of the grasshopper is relatively easy to see. Carefully remove all of the digestive and reproductive organs from the body cavity. Take care to not disturb the thin layer of tissue lining the floor of the body cavity. Use a magnifying lens or dissection scope to locate the whitish to translucent structures of the ventral nerve cord, segmental ganglia of the abdomen, and many radiating nerves of the thoracic ganglia. DRAW what you see.

Group questions

1) For many years taxonomists concluded that lopopods had a body plan that was midway between annelids and arthropods, and that annelids, lobopods, and arthropods were closely related. Recent phylogenetic work has shown that this is not the case. Based on what you saw, tell why you think taxonomists made their original, though incorrect, assumption.
2) How has the specialization of appendages contributed to the success of arthropods?
3) What do you think keeps Hexapods from becoming dominant members of aquatic (especially marine) communities?
4) Horseshoe crabs spend most of their time on soft sediments where they move slowly along searching for small invertebrates that they dig up, crush, and ingest. In what ways is the horseshoe crab body plan well adapted to life on soft substrates?
5) Hexapoda have the most successful body plan of any animal group (at least by number of species). Based on what you've seen, why do you think this is so?

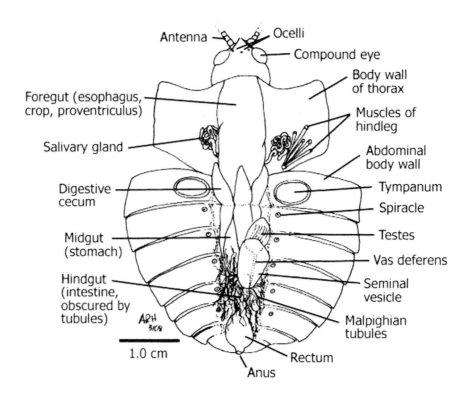

Figure 9.13. Internal anatomy of the grasshopper *Lubber*.

Chapter 10: Echinodermata

Echinodermata belong to Clade Deuterostomia. Echinoderms, chordates, hemichordates, and a few small groups not addressed in this manual are deuterostomes. This chapter focuses on the echinoderms, and next chapter presents hemichordates and chordates.

Traits shared by deuterostomes include:

1) Radial cleavage
2) Blastopore gives rise to the anus
3) Mesoderm formation usually by enterocoely
4) Body divided into three distinct regions: protosome, mesosome, and metasome, each with a pair of coeloms.

In this lab you focus on the Echinodermata, a strictly marine taxon of about 7,000 living species divided into two main groups: the Crinoidea, which includes feather stars and sea lilies (not covered in this lab); and the Eleutherozoa, which in turn includes the Asteroidea, sea stars, Echinoidea, sea urchins and sand dollars, Holothuroidea, sea cucumbers, and Ophiuroidea, brittle stars. The name echinoderm means "spiny skin". Indeed, many members of this group produce spines.

Traits shared by echinoderms include:

1) Water vascular system
2) Connective tissue that can change consistency very quickly (mutable)
3) Secondary pentaradial symmetry (in most)

Asteroidea

Sea stars are important members of most marine environments. They are, in fact, usually top predators in rocky intertidal communities. Sea stars have a global distribution and are members of benthic communities at all depths and all latitudes. Most people are familiar with sea stars because they are relatively easy to spot.

Tasks

1) Immerse a preserved specimen. Observe the external anatomy of your specimen. Be sure to note the central disc and five arms. Identify oral and aboral surfaces of your specimen, and indicate the button-like madreporite located on the aboral surface of the central disc at the junction of two of the arms. Also notice the many podia of tube feet located in the ambulacral grooves of the arms. Finally, take a good look at the mouth region, including oral spines. DRAW what you see.
2) Use a pair of scissors to cut a small piece (1 cm x 1 cm is big enough) from the aboral body wall of one of the arms of your specimen. Immerse that piece of the body wall. Use a dissection scope to examine structures of the body wall. DRAW what you see. Refer to Figure 10.1 to help you identify what you see.
3) Use a pair of fine-tipped forceps to remove several pedicellaria from the body wall. Make a wet mount slide of those structures. Use a compound scope to examine the anatomy of

pedicellaria. DRAW what you see. See if you can deduce how sea star pedicellaria work. Refer to Figure 10.2 as needed.

4) Open the body cavity by using a pair of scissors to make a shallow cut around the edge of the central disc. Cut around the madreporite, thus leaving that structure intact and in place. Use a probe to separate any soft tissues from the body wall as you carefully lift the body wall off of the central disc. Next, use a pair of scissors to cut away the aboral body wall from three arms of your specimen (it doesn't matter which three). Identify structures of the digestive system in the central disc and one of the arms. DRAW the digestive system of the sea star. Look at Figure 10.3 to help you identify what you see.

5) Remove the pyloric cecum from one of the arms of your specimen. Locate the pair of gonads that lie below the pyloric cecum and along the edges the body wall. Also, with pyloric cecum out of the way, gently pull on the cardiac stomach and look for the nearly transparent cardiac stomach retractor muscles that are attached to the ambulacral ridge. DRAW a ray showing the location of the gonads and cardiac stomach retractor muscles. Refer to Figure 10.3 to help you identify these structures.

6) Water vascular system. Remove the pyloric and cardiac stomachs from the central disc, being sure to keep the madreporite and stone canal in place and intact. DRAW the water vascular system. Refer to Figure 10.4 to help you identify what you see.

7) Use a dissection or compound scope to examine a prepared cross section slide of the arm of a sea star. DRAW what you see. Look at Figure 10.5 to help you identify what you see.

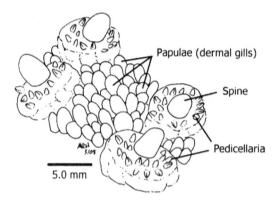

Figure 10.1. Body wall of a sea star.

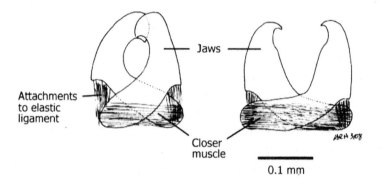

Figure 10.2. Sea star pedicellaria: closed, left; and gaping, right.

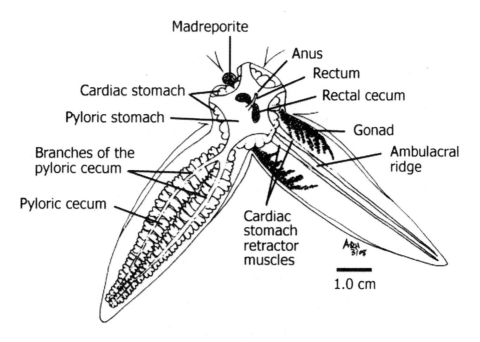

Figure 10.3. Internal anatomy of the central disc and digestive and reproductive systems of a sea star.

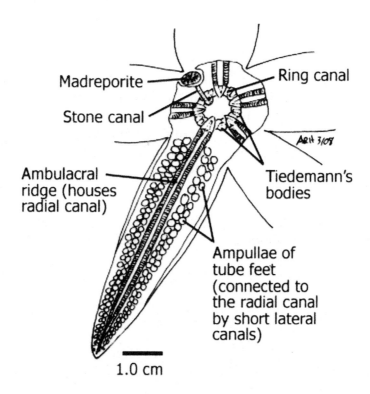

Figure 10.4. Anatomy of the water vascular system of a sea star.

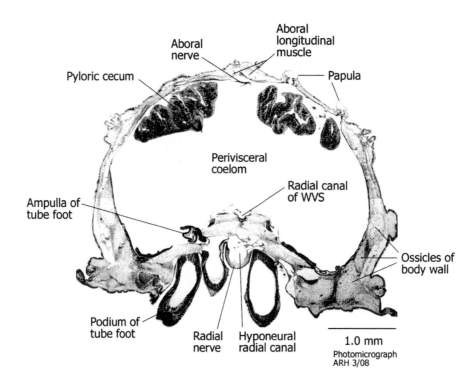

Figure 10.5. Cross-section of the arm of a sea star.

Echinoidea

Sea urchins and sand dollars are echinoids. Sea urchins play an important role in nutrient cycling and community structure of low intertidal and shallow subtidal habitats. Most urchins are opportunistic herbivores that prefer to feed on drift (dead) kelp. Where predators keep urchin populations in check, drift kelp meets the needs of the urchin population and kelp forests or kelp beds can thrive. If, however, urchin populations get too large they consume drift kelp and live kelp, thus decimating kelp forests and kelp beds. Sea otters are the most effective predators at managing sea urchin populations.

Echinoids are characterized by having rigid endoskeletons called tests that are made of hexagonal $CaCO_3$ plates that are fused together. Additional $CaCO_3$ is added to the outer edges of each plate at the echinoid grows.

<u>Tasks</u>

Sea urchin

1) Obtain a live sea urchin, if available. Put it in a large glass bowl filled with seawater or Instant Ocean™. Observe how it uses its tube-feet to move. Flip it upside down and observe how it rights itself. Touch one of its spines and see how it reacts to physical contact. RECORD your observations.
2) Urchin test. Obtain a cleaned urchin test (i.e., shell). The test is made of many hexagonal $CaCO_3$ ossicles that are fused together. Some ossicles have many small holes passing through them, others do not. The holes allow podia of the water vascular system which remain outside of the test to remain connected to the rest of the water vascular system

inside the test. Two hole-bearing ossicles alternate with two non-hole bearing ossicles. There are five sets of hole-bearing ossicles and five sets of non-hole bearing ossicles. The hole-bearing ossicles are referred to as ambulacral plates, while the non-hole bearing ossicles are interambulacral plates. DRAW a set of ambulacral and interambulacral plates. Refer to Figure 10.6 to identify those structures.

3) Use a magnifying glass to observe the oral and aboral poles of your specimen. Structures of the aboral end include the periproct, the madreporite, four genital plates, five ocular plates, and five gonopores. These are sometimes difficult to identify. Refer to Figure 10.6 to help you identify structures of the periproct. Examine the oral pole of the specimen. Identify the five teeth and peristomial membrane surrounding the mouth. DRAW the oral and aboral poles.

4) Use a pair of heavy scissors to penetrate the test, anyplace on the aboral half of the test is OK, and remove a piece of the test that is about 2 cm x 2 cm in size. Immerse your specimen and set it aside. Immerse the small piece of the test and use a dissection scope to observe structures associated with the outer surface of the test. Be sure to observe the relationship between a spine, its muscles, and tubercle. DRAW a spine/tubercle/muscle complex.

5) Next, locate pedicellaria. They are scattered across the surface of the test between the spines. Use a pair of fine-tipped forceps to remove several pedicellaria and make a wet-mount slide of them. Use a dissection or compound scope to study their anatomy. DRAW a pedicellarium from your specimen. Refer to Figure 10.7 to help you identify what you see and discern how it works.

6) Open the body cavity. Use a pair of heavy scissors to cut all the way around the test, just above the equator of the test. Hold the oral and aboral halves of the test together until you complete the cut. The test may fall apart as you make this cut. Use a probe to separate mesenteries that attach the gonads and structures of the digestive system from the inner wall of the aboral half of the test, and then remove the aboral half of the test. All soft tissues of the internal anatomy should be sitting in the oral half of the test. Gently rinse your specimen, change the water in your glass bowl, and re-immerse your specimen.

7) Reproductive system. An aboral view of your specimen should reveal five gonads that can fill most of the space inside the aboral half of the body cavity, i.e., perivisceral coelom, depending on gender and reproductive state of your specimen. Gonads are located along the inner surfaces of the interambulacral plates. Each gonad has its own gonopore in the aboral plate. DRAW what you can see of the reproductive system. Refer to Figure 10. 8 to help you identify what you see.

8) Digestive system. Carefully remove the gonads and rinse your specimen again. Most of the digestive tract is located in the oral half of the perivisceral coelom. The digestive tract is held in place by mesenteries that are connected to the inner wall of the test close near the test's equator. DRAW the digestive tract. Look at Figure 10.8 to help you identify what you see.

9) Water vascular system. The water vascular system of urchins is much more difficult to see than in sea stars, notice the rows of ampullae that line the inner surfaces of ambulacral plates of the test. The ring canal and radial canals can sometimes be seen at the top of the Aristotle's lantern. You should at least be able to see part of the stone canal attached to the axial organ.

10) Aristotle's Lantern. Remove the gonads and the digestive tract. Break away all of the test except for the part that supports Aristotle's lantern. Rinse and then immerse Aristotle's lantern. Examine it using a dissection scope or magnifying lens. Locate the structures indicated on Figure 10.9. Experiment a little by tugging gently on different part of the Aristotle's lantern to see how the different parts move relative to each other. Feel free to disassemble the Aristotle's lantern as you examine it. Enter comments in your lab notebook about what you saw, and about how you think it works. You do not need to draw it unless you REALLY want to...it is cool though!

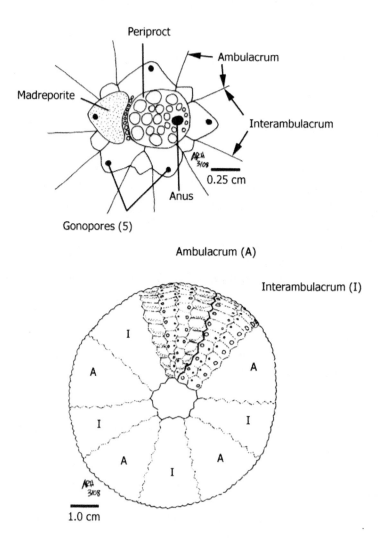

Figure 10.6. Test of a sea urchin, dorsal view, below; detail of the periproct, above.

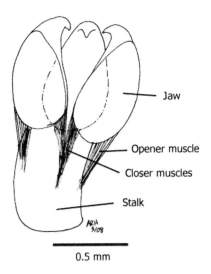

Figure 10.7. Detail of a pedicellarium from a sea urchin.

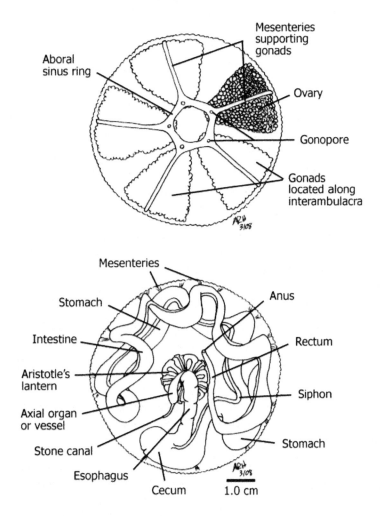

Figure 10.8. Internal anatomy of the sea urchin: reproductive system, above; digestive system, below.

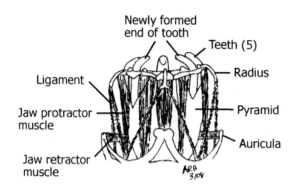
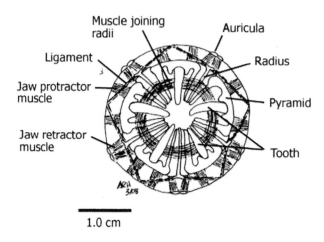

Figure 10.9. Aristotle's lantern: lateral view, above; aboral view, below.

Sand dollars

Sand dollars have dorso-ventrally flattened bodies. Short spines cover the entire body. These animals make a living by capturing particles that are suspended in the water. They do this cooperatively. That is, sand dollars usually live together in large numbers. When a current flows over them they reorient their bodies so that one edge of the body is embedded in the sand and the other edge sticks up above the surface of the sand. This produces back eddies behind and between their tests, water flow rate there slows, and small particles are captured and moved to the mouth.

<u>Tasks</u>

1) Obtain a preserved specimen and study its external anatomy. Look for spines, and, based on what you can see, compare and contrast the spines of sand dollars with those of sea urchins. RECORD your observations.
2) Obtain a cleaned sand dollar test. Study the oral and aboral surfaces of the test. DRAW what you see. Refer to Figure 10.10 to help you identify what you see.

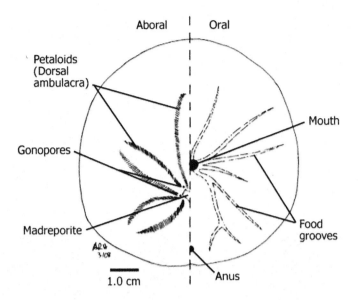

Figure 10.10. External anatomy of the west coast sand dollar *Dendraster excentricus*: aboral view, left, oral view, right.

Holothuroidea

Sea cucumbers present another variation on the echinoderm body plan. These animals have pentaradial symmetry, but in a longitudinal rather than a vertical axis. Sea cucumbers make a living by using their tentacles for suspension or deposit feeding. If you have a chance to observe living sea cucumbers you will observe that they have to physically stick their tentacles in their mouth and lick them as they feed.

Tasks

1) Obtain a preserved sea cucumber and immerse it. Study its external anatomy. DRAW what you see. Refer to Figure 10.11 to help you identify what you see.

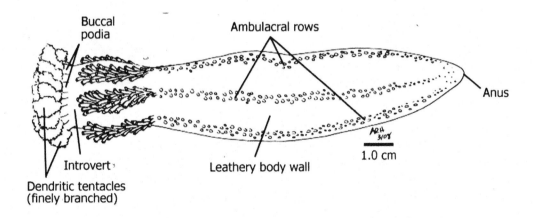

Figure 10.11. External anatomy of the sea cucumber *Cucumaria miniata*.

Echinoderm development

Echinoderm development has been studied extensively because they, like we, are deuterostomes. We can therefore learn a great deal about our own early development by studying theirs.

Tasks

1) Use a compound microscope to examine prepared slides of developmental stages of echinoderms.
2) Develop and label a DRAWING that shows all developmental stages available to you. Refer to Figure 10.12 to help you identify the developmental stages.

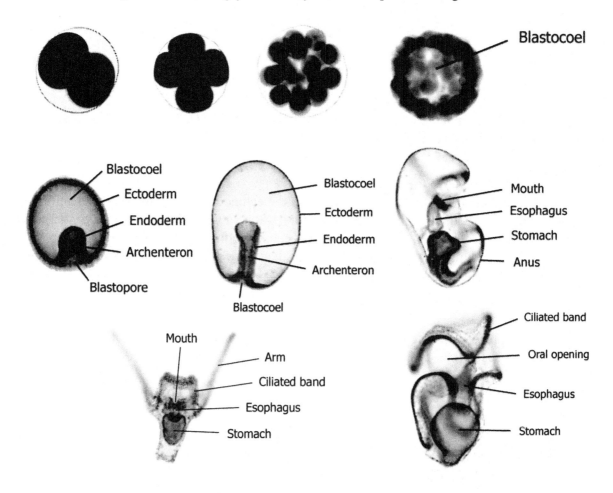

Figure 10.12. Embryonic and larval stages of echinoderms: 2-cell stage with fertilization membrane, top row, left; 4-cell stage with fertilization membrane, top row left-middle; morula stage with fertilization membrane, top row right-middle; blastula stage, top row, right; early gastrula stage, middle row, left; late gastrula stage, middle row, middle; early bipinnaria larva, middle row, right; sea urchin two-arm pluteus stage, bottom row, left; sea star later stage bipinnaria larva, bottom row, right.

Group Questions

1) What evidence is there that all echinoderms have bilateral bodies?
2) Explain why YOU think that radial symmetry is a favorable adaptation for a slow moving, benthic, chemosensory hunters, like sea stars.
3) It's not sharks, killer whales, coral, or any kind of fish that produces the largest number of marine animal-caused injuries to humans each year. Sea urchins do. Based on what you saw today, why do you think this is so?

Chapter 11: Hemichordata and Chordata

Clade Hemichordata is a small group of about 120 species animals that are generally accepted as the sister taxon to Clade Chordata. Hemichordates are assigned to two taxa: Enteropneusta, the deposit-feeding acorn worms; and Pterobranchia, the suspension-feeding pterobranchs.

Hemichordates have the following traits:

1) Tripartite body plan, as described in Chapter 10
2) Stomochord – a diverticulum of the foregut that is a stiff supporting rod that was once thought to be homologous to the notochord of chordates, but is not.

The other taxon in this chapter is Clade Chordata, which includes the Tunicata (formerly called Urochordata), a group of about 2,100 species of sea squirts and their relations; and Metameria, which includes around 73,000 species of chordates that express segmentation, and include the Cephalochordata and Craniata that includes all Vertebrates.

Chordates have the following traits:

1) Dorsal hollow nerve cord
2) Notochord
3) Pharyngeal openings
4) Muscular post-anal tail

Cephalochordates, commonly referred to as lancelets, are studied extensively because they belong to the sister taxon of vertebrates, and, as such, provide clues to the origin of the vertebrate body plan.

Hemichordata

This exercise addresses only the acorn worms. These worms are strictly marine, and live in sand and mud environments where they are deposit feeders.

Tasks

1) Obtain a preserved specimen, rinse it off, and immerse it. Note: Yellowish fluid, a brominated substance produced by the worm, may diffuse out of your specimen along with lots of mucus. Use a magnifying lens or dissection scope to examine the external anatomy of your specimen. DRAW what you see. Refer to Figure 11.1 to help you identify what you see.

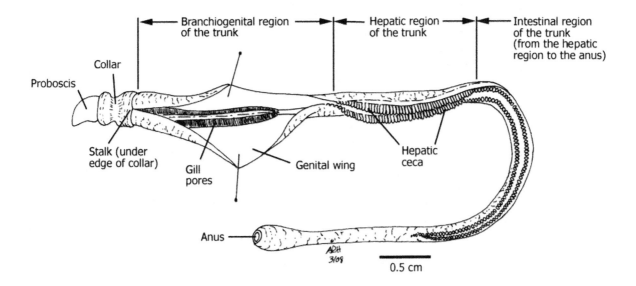

Figure 11.1. External anatomy of the enteropneust acorn worm, *Balanoglossus*. The genital wings are pinned back to reveal the gill pores.

Tunicata (Urochordata)

Tunicates do something extremely bizarre for an animal. They secrete a cellulose-based material called tunicin. Tunicin is the main component of the material in the tunic, their tough protective outer covering. This group contains the following taxa: Ascidiacea, the sea squirts; Thaliacea, salps and doliolarians; and Larvacea, well, the larvaceans. This exercise focuses on two species of ascidians. The first is *Molgula*, a solitary ascidian, and the second is *Ecteinascidia*, a colony-forming species.

Tasks

1) Obtain a specimen of *Molgula* and immerse it. This tunicate lives attached to just about any hard substrate. It prefers the calm waters of harbors, bays, and estuaries, and is found in calm-water marine and estuarine habitats around the world. Carefully remove debris and sediment that normally covers its tunic. Use a magnifying glass or dissection scope to examine your specimen. Some internal anatomy can be seen through the translucent tunic. DRAW what you see, and refer to Figure 11.2 to help you identify what you see.
2) Obtain a prepared slide of the colony-forming ascidian, *Ecteinascidia*. This sea squirt lives in tropical waters, especially in mangrove communities where they live attached to submerged mangrove roots. The tunic of *Ecteinascidia* is transparent, so its anatomy is easily studied. Use a dissection scope or compound scope to study the anatomy of *Ecteinascidia*. Keep in mind as you observe this specimen that the zooid has been flattened to create the slides you are studying, so try to imagine what the specimens would have looked like in their three-dimensional form. DRAW what you see. Refer to Figure 11.3 to help you identify what you see.
3) Obtain a prepared slide of an ascidian tadpole larva. Ascidians produce a swimming, non-feeding larval stage called the tadpole larva. This larva got its name due to its superficial resemblance to a frog tadpole larva. Use a compound scope to study the anatomy of this

larva. The tadpole larva is the only life stage of an ascidian when all characteristics of Clade Chordata are present. DRAW what you see and refer to Figure 11.4 to help you identify what you see.

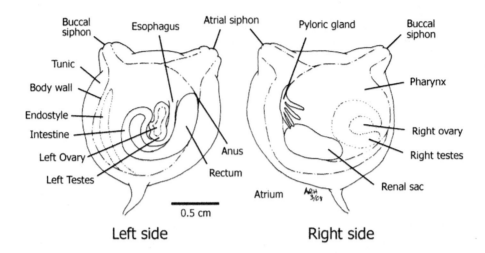

Figure 11.2. External anatomy of *Molgula*, the sea grape.

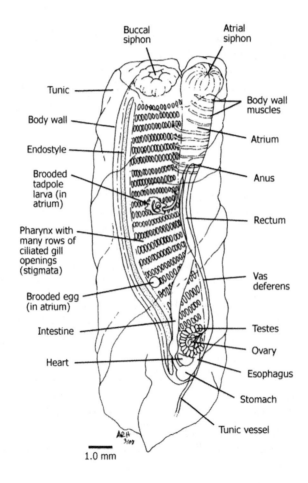

Figure 11.3. Anatomy of one zooid of the colonial ascidian *Ecteinascidia*.

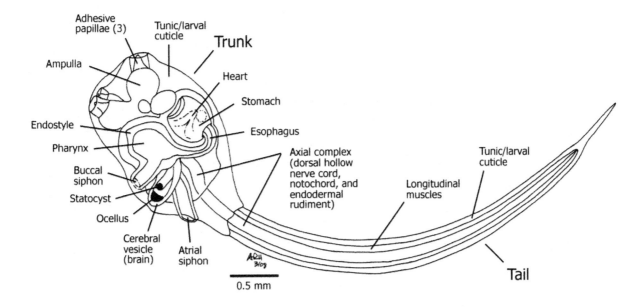

Figure 11.4. Anatomy of a tunicate tadpole larva.

Cephalochordata

Branchiostoma is the representative cephalochordate you will study in this exercise. *Branchiostoma*, like all cephalochordates, is strictly marine and makes its living as a suspension feeder. Cephalochordates live in shallow water tropical and temperate clean sand habitats where they burrow tail-first into the sediment. Only their heads protrude above the sediment surface so they can pull water into their mouths as they filter-feed.

Cephalochordates are good invertebrate representatives of the Clade Metameria. Their metamerism is evident in the repeated muscle blocks and serial gonads.

Tasks

1) Obtain a preserved specimen of *Branchiostoma*. Use a magnifying lens or dissection scope to look for the metameric chevron-shaped muscle blocks that line the flanks of these animals. Also, look for the cream colored gonads that are found along the ventral half of the body.
2) Obtain a prepared whole mount slide of *Branchiostoma* and use a compound scope to study its internal anatomy. DRAW what you see, and refer to Figure 11.5 to help you identify what you see. Take particular care to identify anatomical structures that qualify this animal to be a member of Clade Chordata.
3) Before you return the prepared slide, take a close look at dorsal hollow nerve cord and the many cup ocelli that are found along its edges. Note orientation of the openings into these cup ocelli.
4) Obtain a prepared cross-section slide through the pharyngeal region of *Branchiostoma*. Use a compound scope to study the slide, and DRAW what you see. Look at Figure 11.6 to help you identify what you see.

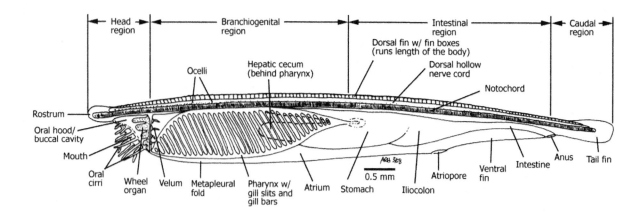

Figure 11.5. Lateral view of *Branchiostoma*.

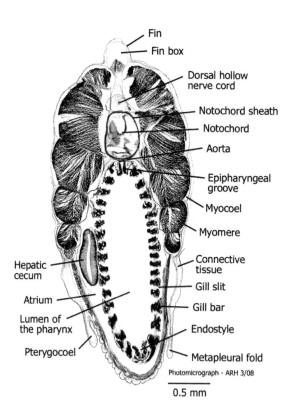

Figure 11.6. Cross-section through the pharynx of *Branchiostoma*.

Group Questions

1) The body ascidian tunicates is often described as a sac-within-a-sac body plan. How would you describe the body plan of cephalochordates? While you are thinking about it, how would you describe the body plan of humans?
2) Do you think that the cephalochordate body plan is ancestral or derived compared to what you would expect the ancestral chordate species to have looked like?

Reference Material

Bogtish, BJ, and TC Cheng. 1998. Human Parasitology. 2nd ed. San Diego, CA: Academic Press. 484 p.

Bullough, WS. 1958. Practical Invertebrate Anatomy. 2nd Ed. London, England: MacMillan. 483 p.

Fox, R. 2001. Invertebrate Anatomy OnLine. *Molgula*. Lander University. http://webs.lander.edu/rsfox/invertebrates/molgula.html

Fox, R. 2005. Invertebrate Anatomy OnLine. *Ecteinascidia*. Lander University. http://webs.lander.edu/rsfox/invertebrates/ecteinascidia.html

Fox, R. 2007. Invertebrate Anatomy OnLine. *Nereis viriens*. Lander University. http://webs.lander.edu/rsfox/invertebrates/nereis.html

Harmer, SF, and AE Shipley, eds. 1910. Worms, Rotifers, and Polyzoa. The Cambridge Natural History, Vol. II. London, Macmillan & Co. Ltd. 560 p.

Harmer, SF, and AE Shipley, eds. 1920. Crustacea and Arachnids. The Cambridge Natural History, Vol. IV. London, Macmillan & Co. Ltd. 566 p.

Harmer, SF, and AE Shipley, eds. 1922. Peripatus, Myriapods, Insects Part I. The Cambridge Natural History, Vol. V. London, Macmillan & Co. Ltd. 584 p.

Meinkoth, NA. 1981. The Audubon Society Field Guide to North American Seashore Creatures. Alfred A. Knopf, Inc. 799 p.

Romoser, WS, and JG Stoffolano, Jr. 1998. The Science of Entomology. 4th ed. Boston, MA: WCB McGraw-Hill. 605 p.

Ruppert, EE, RS Fox, and RD Barnes. 2004. Invertebrate Zoology: A Functional Evolutionary Approach. 7th ed. Belmont, CA: Thomson Brooks/Cole. 963 p.

Struck, *et al.* 2007. Annelid phylogeny and the status of Sipuncula and Echiura. BMC Evolutionary Biology 7: 57-68.

Struck, *et al.* 2011. Phylogenomic analyses unravel annelid evolution. Nature 471: 95-98.

Wallace, RL, and WK Taylor. 2002. Invertebrate Zoology: A Laboratory Manual. 6th ed. Upper Saddle, NJ: Prentice Hall. 356 p.

Index

Abdomen, 92-94, 96, 98, 99, 101-105
Abductor muscle, 84
Aboral sinus ring, 113
Accessory nidamental gland, 60, 64
Aciculum, 75
Acetabulum, 47
Acontia, 40
Acorn worm, 118
Adductor muscle, 66, 68-70, 82-84
Adductor muscle scar, 68
Adhesive papillae, 121
Adjustor muscle, 82-84
Albumin gland, 57, 58
Ambulacral groove, 107
Ambulacral plate, 111, 112, 115
Ambulacral ridge, 108, 109
Ampullae, 109-111, 121
Annelid, 72
Antennae, 75, 90, 96, 97, 100, 102-104, 106
Antennal gland, 99
Antennal muscle, 97, 99
Antennule, 96, 100
Anterior mantle vein, 62
Anthopleura elegantissima, 32
Anthozoa, 32, 39
Anus, 53, 54, 57, 58, 62, 65, 69, 70, 73, 80, 84, 86, 90, 97-99, 106, 107, 109, 112, 113, 115, 116, 119, 120, 122
Aorta, 52, 54, 102, 122
Aplacophora, 51
Apopyle, 30
Aquiferous pore, 61
Arachnida, 93, 95
Archenteron, 116
Aristotle's lantern, 112-114
Armadillo bug, 96
Arthropod, 51, 81, 85, 90, 92, 104
Ascaris lumbricoides, 85-88
Ascidiacae, 119, 120
Asconoid, 27, 29, 31
Asteroidea, 107
Atrial siphon, 120. 121
Atriopore, 122
Atrium, 120, 122
Aurelia, 36, 37
Auricle, 43
Auricula, 114
Axial complex, 121
Axial furrow, 92
Axial organ, 111, 113

Axial ring, 92
Balanoglossus, 119
Barnacle, 96, 102
Basal lamina, 42
Beak, 60
Beetle, 104
Bipinnaria, 116
Bivalve, 51, 66, 67
Blastocoel, 116
Blastopore, 38, 107, 116
Blastostyle, 35
Blastula, 116
Blue crab, 96, 97, 101
Body column, 39, 40
Book gills, 94
Box jelly, 32, 38
Brain, 42, 47, 60, 73, 75, 98, 101, 121
Branchiogenital region, 119, 122
Brachiopoda, 81, 82, 93
Branchial chamber, 94
Branchial heart, 59, 60, 62
Branchiostoma, 121, 122
Brittle star, 107
Buccal bulb, 54, 58, 60, 66, 76
Buccal cavity, 52, 55, 66, 78, 122
Buccal podia, 115
Buccal retractor muscle, 76
Buccal siphon, 120, 121
Buthus occitanus, 95
Cadherin, 25, 32
Caenorhabditis elegans, 85
Calcarea, 25
Calciferous gland, 78
Callinectes, 96, 101, 102
Calydiscoides euzeti, 45
Carapace, 94, 97, 101, 102, 103
Carapace spine, 97
Cardiac stomach, 99, 101, 108, 109
Cardiac stomach retractor muscle, 108, 109
Cartilaginous groove, 62
Caudal region, 122
Cecum, 45, 47, 62, 63, 65, 76, 113
Cellularia, 25
Centipede, 103, 104
Central disc, 107-109
Cephalization, 42
Cephalon, 92
Cephalochordata, 118, 121
Cephalopod, 51, 59
Cephalothorax, 93, 94, 96, 98, 99, 101-103

Cercaria, 45-47
Cerebral ganglion, 58, 76, 78
Cerebral vesicle, 121
Cestoda, 43, 48
Chaetae, 72, 75, 77-79, 83
Chelicerae, 93-95
Chelicerata, 92, 93
Cheliped, 100, 102
Chilarium, 94
Chilopoda, 103, 104
Chinese liver fluke, 45, 46
Chironex fleckeri, 32
Chitin, 72, 81, 85, 92
Chiton, 51, 52
Chlorogogen cells, 76, 78, 79
Choanocyte, 25, 27, 30
Chordate, 107, 118, 120, 121
Chucky pig, 96
Circular muscle, 72, 76, 79, 105
Circulatory system, closed, 72
Circulatory system, open, 51
Circumesophageal connective nerve, 97, 99
Cirrus, 49, 50
Cirrus pouch, 49, 50
Clade, 18
Cladogram, 19
Clam, 51, 66
Clitellata, 72
Clitellum, 78, 80
Cloaca, 86, 87
Cnidaria, 32
Cnidocil, 34
Cnidocyte, 32-34, 36
Coelom, 51, 72, 73, 74, 76, 79, 82-84, 107
Collagen, 72
Collar, 56, 57, 119
Compensation sac, 73
Compound eye, 92, 93, 97, 99, 102, 103, 106
Connective tissue, 72, 107
Contractile vessel, 72, 73
Copulatory bursa, 58
Copulatory bursa duct, 58
Coral, 32, 39, 40
Coxa, 105
Crab, 96, 101, 102
Craniata, 118
Cranium, 60
Crayfish, 96-101
Crinoid, 107
Crop, 58, 78, 106
Crustacea, 92, 96, 102
Cryptochiton stelleri, 53, 54

Ctenidium, 53, 62, 67, 69
Cubozoa, 32, 38, 39
Cubomedusa, 39
Cucumaria miniata, 115
Cuticle, 72, 76, 79, 81, 83, 85, 87, 88, 90, 96
Cuttlefish, 59
Cycloneuralia, 81
Cysticercus, 48
Daphnia, 96, 97
Dactyl, 102
Dart sac, 58
Demospongia, 25
Dendraster excentricus, 115
Dermal gill, 108
Deuterostomia, 107
Digenea, 42-44, 48
Digestive cecum, 82-84, 97, 99, 101, 102, 106
Digestive gland, 52, 53, 57, 58, 60, 62, 65, 67, 70
Diploblastic, 32
Diplopoda, 103, 104
Doliolaria, 119
Dorsa brood pouch, 97
Dorsal cirrus, 75
Dorsal shield, 92
Dorsal spine, 103
Doublure, 92
Dugesia, 42
Earthworm, 72, 77-79
Ecdysis, 85
Ecdysozoa, 81
Echiniscus spinulosus, 91
Echiniscus testudo, 91
Echinodermata, 107, 116
Echinoidea, 107, 110
Echinoplectanum leave, 45
Ecteinascidia, 119, 120
Ectoderm, 38, 42, 116
Ectoparasite, 43, 44
Efferent branchial vessel, 62
Ejaculatory duct, 87
Eleutherozoa, 107
Embryonated egg, 45-48, 85
Endoderm, 38, 42, 85, 116
Endostyle, 120-122
Enterocoely, 107
Enteropneusta, 118, 119
Ephyra, 36, 38
Epidermis, 33, 42, 43, 72
Epipharyngeal groove, 122
Errantia, 72, 74
Esophagus, 47, 55, 58, 60, 65, 66, 67, 70, 73, 76, 78, 99, 106, 113, 116, 120, 121

Eumetazoa, 25, 32
Euplectella, 26
Eutely, 85
Excretory bladder, 47
Exhalant aperture, 53
Exhalant chamber, 53
Exhalant siphon, 68, 69
Exopod, 97, 101
Exoskeleton, 97, 98, 101, 103, 104
Eyespot, 45
Facial suture, 92
Fang, 104
Fin, 61, 62, 122
Fin box, 122
Flatworm, 42
Flagellum, 58
Fluke, 44, 45
Food groove, 93, 115
Foot, 53, 56, 57, 60, 69
Foot protractor muscle, 68
Foot protractor muscle scar, 68
Foot retractor muscle, 68
Foot retractor muscle scar, 68
Foregut, 106
Forcipule, 104
Funnel, 61, 62
Ganglion, 98, 99, 105
Gastric mill, 98, 101
Gastric muscle, 99, 102
Gastric ostia, 39
Gastric pocket, 37
Gastrodermis, 33, 38
Gastropod, 51, 55
Gastrovascular cavity/Coelenteron, 32-34, 38-43
Gastrozooid, 34, 35
Gastrula, 116
Gemmule, 30, 31
Gena, 92
Genital opening, 56-58
Genital operculum, 94, 95
Genital plate, 111
Genital pore, 47, 49, 50
Genital sensillum, 87
Genital wing, 119
Giant neuron, 72
Gill, 52, 53, 59, 60, 62, 66, 67, 69, 70, 97, 99, 101, 119, 122
Gill bar, 122
Gill chamber, 97, 101, 102
Gill cleaner, 101 102
Girdle, 52-54
Gizzard, 78

Glabella, 92
Glabellar furrow, 92
Glochidia larva, 67, 70
Gnathobase, 94
Gonad, 32, 36, 37, 52, 54, 57, 58, 67, 70, 73, 82-84, 108, 109, 111-113, 121
Gonangium, 35
Gonochoric, 32, 33, 36, 39
Gonopod, 100
Gonopore, 53, 78, 80, 86, 90, 98, 111-113, 115
Gonotheca, 35
Gonozooid, 35
Grantia, 29
Grasshopper, 104, 106
Green gland, 99
Haptor/Anchor, 45
Head organ, 45
Head retractor muscle, 60, 65
Heart, 52, 78, 99, 101-103, 105, 120, 121
Helix, 55-57
Hemichordate, 107, 118
Hemocyanin, 51, 58
Hemoglobin, 72
Hepatic ceca, 119, 122
Hepatic region, 119
Hermatypic, 40
Hexapoda, 92, 104
Hindgut, 106
Hinge, 70, 81
Hinge teeth, 84
Hirudo, 77, 80
Holothuroidea, 107, 115
Horseshoe crab, 93
Hydra, 33, 34
Hydranth, 35
Hydromedusa, 32, 36, 39
Hydrozoa, 32, 33, 36, 39
Hyponeural radial canal, 110
Iliocolon, 122
Incurrent canal, 29, 30
Index fossil, 81, 92
Inhalant aperture, 53
Inhalant chamber, 53
Inhalant siphon, 68, 69
Ink sac, 60, 62
Ink sac vessel, 62
Insect, 104
Interambulacral plate, 111, 112
Intestinal region, 119, 122
Intestine, 43, 47, 48, 52-54, 57, 58, 62, 65, 67, 70, 73, 74, 76, 78, 79, 83, 84, 86-88, 98, 102, 106, 113, 120, 122

Introvert, 72, 73, 115
Introvert retractor muscle, 73
Jaw, 66, 74, 76, 113
Jaw protractor muscle, 114
Jaw retractor muscle, 114
Jellyfish, 32, 36
Kidney, 57, 58, 60, 62
Kinorhyncha, 81
Krill, 96
Labial palp, 67, 69
Labrum, 103
Lappet, 37, 38
Larvacea, 119
Lateral epidermal cord, 85-88
Lateral excretory vessel, 47
Lateral eye, 95
Lateral mantle artery, 62
Lateral spine, 94
Laurer's canal, 47
Leech, 72, 77, 80
Leucosolenia, 27
Leuconoid, 30, 31
Limulus, 93, 94
Lingula, 66
Lingula, 81, 83, 93
Lobopodia, 81, 90
Lobster, 96
Longitudinal muscle, 72, 76, 79, 87, 88, 105, 110, 121
Lophophore, 81-84
Lophotrochozoa, 81
Loricifera, 81
Lubber, 106
Lumbricus, 77-79
Madreporite, 107-109, 111, 112, 115
Malpighian tubules, 103, 106
Mandible, 66, 96, 100, 103
Mandibular muscle, 99, 100, 102
Mantle, 51, 55, 59-61, 66, 67, 68, 70, 82
Mantle cartilage, 62
Mantle cavity, 51-53, 55-57, 59, 60, 70, 82
Mantle, skirt, 68-70
Manubrium, 34-39
Manus, 102
Maxillae, 96, 100
Maxilliped, 96, 100
Medial mantle artery, 59, 62
Medial mantle vein, 62
Medial septum, 59
Median eye, 95
Medusa, 32, 33, 35-39
Megalops, 102, 103

Mehli's gland, 47, 49, 50
Mesoderm, 42, 85, 107
Mesoglea, 32, 33
Mesohyle, 30
Mesosoma, 95, 107
Metacercaria, 45-47
Metameria, 118, 121
Metamerism, 72
Metanephridium, 51, 72-74, 78, 79
Metapleural fold, 122
Metasoma, 95, 107
Metazoa, 25
Metridium, 39, 40
Microscope, compound, 12-13
Microscope, dissection, 14-15
Midgut, 106
Millipede, 103, 104
Miricidium, 45, 46
Molgula, 119, 120
Mollusc, 51
Monogenea, 42-45
Monoplacophora, 51, 59
Morula, 116
Moss piglet, 91
Mucous gland, 58
Muscle block, 121
Muscular foot, 51
Mussel, 51, 66, 70
Myocoel, 122
Myomere, 122
Myriapoda, 92, 103
Naupliar eye, 103
Nauplius larva, 96, 102, 103
Nephridial canal, 49, 50
Nephridial sac, 67
Nematocyst, 32, 34
Nematoda, 81, 84, 85
Nematomorpha, 81
Neoblast, 42
Neodermata, 42, 43
Neodermis, 43
Nephridiopore, 47, 53, 73, 79
Nephridium, 54, 69, 70
Nereis, 74-77
Nerve cord, 42, 73, 74, 76, 79, 87, 88, 98, 99, 101, 105, 110, 181, 121, 122
Nerve ring, 58
Neuropodium, 75
Nidamental gland, 59, 60, 64
Notochord, 118, 121, 122
Notochord sheath, 122
Notopodium, 75

Nuchal organ, 72, 75
Obelia, 33, 34, 36, 38
Ocellus, 42, 43, 75, 80, 106, 121, 122
Octopus, 51, 59
Ocular plate, 111
Odontophore, 52, 55, 66
Olfactory crest, 61, 62
Olfactory groove, 61, 62
Oncosphere, 48
Onychophora, 81, 90
Ophiuroidea, 107
Opisthorchis sinenesis, 45-47
Oral arm, 37, 38
Oral cirri, 122
Oral disc, 40
Oral hood, 122
Oral opening, 116
Oral papilla, 90
Oral spine, 107
Oral sucker, 47
Oral surface, 107
Organs of Verrill, 63
Oscula/osculum, 27-29
Ossicle, 110
Ostia/ostium, 27, 29, 30, 70, 99, 102
Ovary, 45, 47, 49, 64, 78, 86, 101, 102, 113, 120
Oviduct, 56, 58-60, 64, 78, 86, 87
Oviductal gland, 60, 64
Ovotestis, 57, 58
Pallial line, 68
Palp, 75, 76
Panarthropoda, 81, 90
Pancreas, 65
Papulae, 108, 110
Parapodium, 72, 74-76
Parasite, 42-44, 48, 84
Peanut worm, 72
Pedal disc, 34, 38, 40
Pedicellarium, 107, 108, 111, 113
Pedicle, 35, 81, 83, 84
Pedipalp, 93, 95
Peduncle gland, 45
Pelagosphera larva, 72
Pen, 61, 62
Penis, 58, 59, 62, 63
Penis retractor muscle, 58
Pereopod, 96, 98, 100, 102, 103
Pericardium, 57, 67, 69, 70
Periderm, 35
Periostracum, 67, 68
Periproct, 111, 112
Peristomial membrane, 111

Peristomium, 72, 75, 78
Peritoneum, 72, 85
Perivisceral coelom, 110, 111
Petaloid, 115
Pharyngeal cavity, 43
Pharyngeal muscle, 76, 78
Pharyngeal opening, 118
Pharynx, 38-45, 47, 74, 76, 78, 86-88, 120-122
Phototaxis, 42
Physalia, 33, 36
Pill bug, 96
Placiphorella velata, 51
Planaria, 42, 43
Planula larva, 32, 36, 38
Platyhelminthes, 41
Pleopod, 96, 98, 100, 101, 103
Pleural lobe, 92
Pleurite, 92
Pluteus larva, 116
Pneumostome, 55, 56
Podium, 107, 110
Polychaeta, 72
Polyp, 32, 33, 36, 38, 39
Polyplacophora, 51
Porifera, 25
Portuguese man-of-war, 33
Post-abdomen, 92, 93, 97
Post-abdominal claw, 97
Potato bug, 96
Priapula, 81
Proboscis, 119
Proglottid, 48, 49
Pronatum, 105
Prosoma, 93, 95
Prostomial nerve, 78
Prostomium, 72, 75, 77, 78, 80
Protonephridia, 42
Protosome, 107
Proventriculus, 106
Pseudocoelom, 85-88
Pterobranchia, 118
Pterygocoel, 122
Pulmonary plexus, 57
Pusher leg, 94
Pycnogonid, 93
Pygidium, 72, 92
Pyloric cecum, 108-110
Pyloric gland, 120
Pyloric stomach, 99, 101, 108, 109
Pyramid, 114
Radial canal, 26, 37, 109-111
Radial cleavage, 107

Radial fixing muscle, 73
Radial nerve, 110
Radius, 114
Radula, 51, 52, 54, 55, 60, 66
Radular sac, 52, 54, 55, 66
Rectal cecum, 109
Rectum, 57, 58, 62, 65, 70, 73, 106, 109, 113, 120
Redia, 45-47
Renal sac, 62
Resilium, 68
Rhopalium, 37-39
Ring canal, 37, 109, 111
Rolly-polly bug, 96
Rostellar hooks, 49
Rostellum, 49
Rostrum, 97-99, 101-103, 122
Roundworm, 81, 84
Saccate nephridia, 96
Salivary gland, 52, 54, 58, 65, 66, 106
Salp, 119
Sand dollar, 107, 110, 114, 115
Scale bar, 9-11
Scaphopod, 51
Scolex, 48, 49
Scolopendra, 104
Scorpion, 93, 94
Scypha, 29, 30
Scyphistoma, 36, 38
Scyphozoa, 32, 36, 38, 39
Sea anemone, 32, 39
Sea cucumber, 107, 115
Sea spider, 93
Sea squirt, 118, 119
Sea star, 107-110
Sea urchin, 107, 110, 112, 113
Sea wasp, 32
Sedenteria, 72, 77
Segment, 72, 77, 78, 80, 90, 92, 103, 104, 118
Seminal receptacle, 47, 78, 101, 102
Seminal vesicle, 78, 87, 106
Septal funnel, 37
Septum, 39-41, 74, 76, 78
Setae, 101, 103
Shell, 51, 52, 54-56, 66, 68-70, 81-83
Shell gland, 47
Shell hinge, 66
Shell ligament, 67, 68
Shrimp, 96
Siphon, 60-62, 113
Siphon retractor muscle, 60, 62
Siphonophore, 33
Sipuncula, 72, 73

Sipunculus nudis, 73
Slug, 51, 55
Snail, 51, 55
Solenogaster, 51
Sow bug, 96
Sperm bulb, 63
Sperm duct, 49, 50, 78, 87, 88
Sperm groove, 78
Spermatophore, 101
Spermatophoric gland, 59, 62, 63
Spermatophoric sac, 62, 63
Spicule, 25, 28
Spider, 93, 95
Spindle muscle, 73
Spiracle, 90, 106
Spirobolus, 104
Sponge, 25
Spongin, 30, 31
Spongocoel, 29, 30
Sporocyst, 45, 46
Squid, 51, 59, 61-66
Starfish (see Sea star)
Statocyst, 100, 121
Stellate ganglion, 62
Sternite, 103
Sternum, 95, 101, 102
Stigmata, 95
Stomach, 52-54, 58, 62, 63, 65, 67, 70, 73, 83, 84, 98, 99, 101, 102, 106, 113, 116, 120-122
Stomochord, 118
Stone canal, 108, 109, 111, 113
Strobila, 36, 38
Strobilae, 49
Subradular sac, 55, 66
Sucker, 47, 49, 59, 60, 78
Swimmerette, 96, 98
Syconoid, 29, 31
Symmetry, bilateral, 42
Symmetry, pentaradial, 107, 115
Symmetry, radial, 32
Symplasma, 25
Syncytium, 42
Systemic heart, 60, 63
Tadpole larva, ascidian, 119-121
Taenia saginata, 48, 49
Tagma, 92, 93, 96, 104
Tail, muscular post-anal, 118
Tail spine, 94
Tapeworm, 42, 48
Tardigrada, 81, 90, 91
Taxon, 18
Teeth, cardinal, 68

Teeth, lateral, 68
Tegument, 43
Telson, 94, 98, 99
Tensilium, 67, 68, 69
Tentacle, 32-34, 36, 37, 39, 40, 56-61, 81, 83, 84, 115
Tentacular cirri, 75
Terebratalia, 81, 82, 84
Tergite, 103
Test, 110, 112
Testis/testes, 45, 47, 49, 62, 63, 87, 88, 99, 101, 106
Thaliacea, 119
Theca, 35
Thoracopods, 97
Thorax, 92, 104-106
Tick, 93, 95
Tiedemann body, 109
Tomosvary organ, 103
Tongue, 90
Totipotent, 25
Trachea, 90, 103
Trilobite, 92
Trilobite larva, 93, 94
Tripartite brain, 96
Tripedalia cystophora, 39
Triploblastic, 42, 51
Trochophore, 51, 72
Trunk, 80, 97, 121
Tubatrix aceti, 85
Tube feet, 107, 110
Tubeworm, 72
Tunic, 119, 120, 121
Tunic vessel, 120
Tunicata, 118, 119, 121
Tunicin, 119
Turbellaria, 42, 43

Tusk shell, 51
Tympanum, 106
Typhlosole, 79
Umbo, 67-70
Unio, 66-70
Urochordata, 118, 119
Uropod, 98, 99, 100
Uterus, 47, 49, 50, 85, 86, 87
Vagina, 49, 50, 86
Vas deferens, 47, 58, 62, 87, 88, 99, 106, 120
Vas efferens, 47
Velarium, 39
Velum, 32, 36, 39, 122
Velvet worm, 90
Vena cava, anterior, 62
Vena cava, posterior, 59, 62
Ventral cirrus, 75
Ventricle, 54, 70
Vertebrate, 118
Vinegar eel, 85
Visceral hump, 61
Visceral mass, 51, 56, 59, 60, 67, 69, 70
Vitellarium, 45, 47, 49
Walking leg, 93, 94, 96, 100, 102, 104
Walking worm, 90
Water bear, 90, 91
Water flea, 96, 97
Water vascular system, 107-111
Wet mount slide, 16-17
Wheel organ, 122
Wing, 104, 105
Woodlouse, 96
Xiphosura, 93
Yolk gland, 47
Zoea, 102, 103
Zooid, 32, 33, 35